Edited by
Stavros Kromidas

**The HPLC-MS Handbook
for Practitioners**

Contents

Preface *XI*

The Structure of *HPLC-MS for Practitioners* *XIII*

List of Contributors *XV*

Part I Overview, Pitfalls, Hardware-Requirements *1*

1 **State of the Art in the LC/MS** *3*
O. Schmitz
1.1 Introduction *3*
1.2 Ionization Methods at Atmospheric Pressure *5*
1.2.1 Overview of API Methods *6*
1.2.2 ESI *6*
1.2.3 APCI *8*
1.2.4 APPI *9*
1.2.5 APLI *10*
1.2.6 Determination of Ion Suppression *11*
1.2.7 Best Ionization for Each Question *11*
1.3 Mass Analyzer *12*
1.4 Future Developments *13*
1.5 What Should You Look for When Buying a Mass Spectrometer? *14*
References *15*

2 **Technical Aspects and Pitfalls of LC/MS Hyphenation** *19*
M.M. Martin
2.1 Instrumental Requirements for LC/MS Analysis – Configuring the Right System for Your Analytical Challenge *20*
2.1.1 (U)HPLC and Mass Spectrometry – Not Just a Mere Front-End *20*
2.1.2 UHPLC System Optimization – Gradient Delay and Extra-column Volumes *21*
2.1.3 Does Your Mass Spectrometer Fit Your Purpose? *33*

2.1.4	Data Rates and Cycle Times of Modern Mass Spectrometers *38*
2.1.5	Complementary Information by Additional Detectors or Mass Spectrometry Won't Save the World *39*
2.2	LC/MS Method Development and HPLC Method Adaptation – How to Make My LC Fit for MS? *43*
2.2.1	Method Development LC/MS – LC Fits the MS Purposes *44*
2.2.2	Converting Classical HPLC Methods into LC/MS *53*
2.3	Pitfalls and Error Sources – Sometimes Things Do Go Wrong *54*
2.3.1	No Signal at All *54*
2.3.2	Inappropriate Ion Source Settings and Their Impact on the Chromatogram *56*
2.3.3	Ion Suppression *58*
2.3.4	Unknown Mass Signals in the Mass Spectrum *59*
2.3.5	Instrumental Reasons for the Misinterpretation of Mass Spectra *65*
2.4	Conclusion *68*
2.5	Abbreviations *69*
	References *70*

3	**Aspects of the Development of Methods in LC/MS Coupling** *73*
	T. Teutenberg, T. Hetzel, C. Portner, S. Wiese, C. vom Eyser, and J. Tuerk
3.1	Introduction *73*
3.2	From Target to Screening Analysis *74*
3.2.1	Target Analysis *74*
3.2.2	Suspected-Target Screening *74*
3.2.3	Non-target Screening *74*
3.2.4	Comparable Overview of the Different Acquisition Modes *75*
3.3	The Optimization of Parameters in Chromatography and Mass Spectrometry *75*
3.3.1	Requirements and Recommendations for HPLC/MS Analysis Taking DIN 38407-47 as an Example *75*
3.3.2	The Definition of Critical Peak Pairs in the Context of HPLC/MS Coupling *77*
3.3.3	The Separation of Polar Components from the Column Void Time *79*
3.3.4	Determining the HPLC Method Parameters Using the Example of the Separation of Selected Pharmaceuticals *80*
3.3.5	Carrying out Screening Experiments *84*
3.3.6	Evaluation of the Data and Discussion of the Influencing Parameters *86*
3.3.7	Using Simulation Software for Fine Optimization *98*
3.3.8	Choosing the Stationary Phase Support *99*
3.3.9	The Influence of the Inner Column Diameter and the Mobile Phase Flow Rate *103*
3.3.10	The Influence of the Injection Volume *104*
3.3.11	Establishing the Mass Spectrometric Parameters *115*

Edited by
Stavros Kromidas

The HPLC-MS Handbook for Practitioners

Verlag GmbH & Co. KGaA

Editor

Dr. Stavros Kromidas
Consultant, Saarbrücken
Breslauer Str. 3
66440 Blieskastel
Germany

All books published by Wiley-VCH are carefully produced. Nevertheless, authors, editors, and publisher do not warrant the information contained in these books, including this book, to be free of errors. Readers are advised to keep in mind that statements, data, illustrations, procedural details or other items may inadvertently be inaccurate.

Library of Congress Card No.:
applied for

British Library Cataloguing-in-Publication Data:
A catalogue record for this book is available from the British Library.

Bibliographic information published by the Deutsche Nationalbibliothek
The Deutsche Nationalbibliothek lists this publication in the Deutsche Nationalbibliografie; detailed bibliographic data are available on the Internet at http://dnb.d-nb.de.

© 2017 WILEY-VCH Verlag GmbH & Co. KGaA, Boschstr. 12, 69469 Weinheim, Germany

All rights reserved (including those of translation into other languages). No part of this book may be reproduced in any form – by photoprinting, microfilm, or any other means – nor transmitted or translated into a machine language without written permission from the publishers. Registered names, trademarks, etc. used in this book, even when not specifically marked as such, are not to be considered unprotected by law.

Cover Design   Formgeber, Mannheim, Deutschland
Typesetting   le-tex publishing services GmbH, Leipzig, Deutschland
Printing and Binding CPI Group (UK) Ltd, Croydon, CR0 4YY

Print ISBN 978-3-527-34307-2
ePDF ISBN 978-3-527-80919-6
ePub ISBN 978-3-527-80917-2
Mobi ISBN 978-3-527-80918-9
oBook ISBN 978-3-527-80920-2

Printed on acid-free paper.

C9783527343072_240725

The manufacturer's authorized representative according to the EU General Product Safety Regulation is Wiley-VCH GmbH, Boschstr. 12, 69469 Weinheim, Germany, e-mail: Product_Safety@wiley.com.

3.3.12	Optimization of the Mass Spectrometric Parameters *117*
3.3.13	Quantification Using LC/MS *122*
3.3.14	Screening Using LC/MS *128*
3.3.15	Miniaturization – LC/MS Quo Vadis? *132*
	References *135*

Part II   Tips, Examples, Trends *139*

4   LC/MS for Everybody/for Everything? – LC/MS Tips *141*
F. Mandel

4.1	Introduction *141*
4.2	Tip Number 1 *142*
4.2.1	Choosing the Right LC/MS Interface *142*
4.3	Tip Number 2 *148*
4.3.1	Which Mobile Phases Are Compatible with LC/MS? *148*
4.4	Tip Number 3 *149*
4.4.1	Phosphate Buffer – The Exception *149*
4.5	Tip Number 4 *150*
4.5.1	Paired Ions *150*
4.5.2	Which "Antidote" Is Available? *151*
4.5.3	Summary *152*
4.6	Tip Number 5 *152*
4.6.1	Using Additives to Enhance Electrospray Ionization *152*
4.6.2	Additives for APCI *153*
4.6.3	Summary *154*
4.7	Tip Number 6 *154*
4.7.1	How Can I Enhance Sensitivity of Detection? *154*
4.8	Tip Number 7 *155*
4.8.1	No Linear Response and Poor Dynamic Range? *155*
4.8.2	The Reasons *156*
4.8.3	Possible Solutions *156*
4.8.4	Summary *157*
4.9	Tip Number 8 *157*
4.9.1	How Much MS^n Do I Need? *157*
4.9.2	Solutions *158*
4.9.3	Summary *158*
4.10	Need More Help? *166*
	References *167*

Part III User Reports 169

5 LC Coupled to MS – a User Report 171
A. Muller and A. Hofmann
References 176

6 Problem Solving with HPLC/MS – a Practical View from Practitioners 177
E. Fleischer

- 6.1 Introduction and Scope 177
- 6.2 Case Example 1 181
- 6.2.1 Investigation of Methohexital Impurities and Decomposition Products 181
- 6.2.2 Sample Preparation 181
- 6.3 Case Example 2 183
- 6.3.1 Separation of Oligomers from Caprolactam, Multicomponent Separation of Impurities on a Gram Scale 183
- 6.4 Case Example 3 184
- 6.4.1 Preparation and Isolation of bis-Nalbuphine from Nalbuphine 184
- 6.5 Case Example 4 186
- 6.5.1 Isolation and Elucidation of Dopamine Impurities 186

7 LC/MS from the Perspective of a Maintenance Engineer 189
O. Müller

- 7.1 Introduction and Historical Summary 189
- 7.2 Spray Techniques 190
- 7.3 Passage Through the Ion Path 191
- 7.4 The Analyzer 191
- 7.5 Maintenance 193
- References 198

Part IV Vendor's Reports 201

8 LC/MS – the Past, Present, and Future 203
T.L. Sheehan and F. Mandel

9 Vendor's Report – SCIEX 207
D. Schleuder

10	**Manufacturer Report – Thermo Fisher Scientific** *213*	
	M.M. Martin	
10.1	Liquid Chromatography for LC/MS *214*	
10.2	Mass Spectrometry for LC/MS *215*	
10.3	Integrated LC/MS Solutions *217*	
10.4	Software *217*	
	References *219*	

About the Authors *221*

Index *227*

Preface

LC/MS coupling has developed from a method for experts in research to a well-proven technique for users in their daily routine. Hence, this book is dedicated exclusively to LC/MS coupling.

It is our goal to give LC/MS users detailed information in order to use *their* LS/MS application in an optimal manner. Colleagues who have authored articles in my previous books have therefore revised and updated their articles. Furthermore, new articles from LC/MS practitioners were added. When writing those articles, it was most important to us to have an eye on practice, but compact background knowledge is also given. I hope that the analyst in development as well as the user in daily routine will find inspiration and tips for optimal usage of LC/MS coupling.

My special thanks goes to Wolfgang Dreher for his critical comments to this manuscript, furthermore to my author colleagues, who put down their experience and knowledge in writing despite their limited time resources. I would like to thank WILEY-VCH and in particular Reinhold Weber and Martin Preuss for the good and close cooperation.

Blieskastel, March 2017 *Stavros Kromidas*

The Structure of *HPLC-MS for Practitioners*

The book contains ten chapters that are divided into four parts:

- 1–3: Part I Overview, Pitfalls, Hardware Requirements
- 4: Part II Tips, Examples
- 5–7: Part III User Reports
- 8–10: Part IV Vendor's reports, Trends

Part I

In Chapter 1 Oliver Schmitz overviews the **State of the art of LC/MS coupling** and opposes different modes. In Chapter 2 Markus Martin shows **Technical aspects and pitfalls of LC/MS hyphenation** and provides precise and specific hints how LC/MS coupling can successfully be established in daily routine. Other topics of Chapter 2 are method development as well as method transfer. Thorsten Teutenberg and co-authors provide a great many of suggestions in Chapter 3 (**Requirements of LC hardware for the coupling of different mass spectrometers**), as to arrange LC/MS coupling as optimal as possible. Among other things complex samples and miniaturization play an important role.

Part II

In Chapter 4 Friedrich Mandel offers a numerous of **LC/MS Tips** addressing different topics of LC/MS coupling.

Part III

Chapter 5 contains examples and experience reports from users and service engineers: LC/MS coupling is often linked to life science and environmental analysis. Alban Muller and Andreas Hofmann show in Chapter 5 a **concrete example of LC/MS coupling in ion chromatography** as an unfamiliar application. In Chapter 6 Edmond Fleischer shows on the basis of 4 examples coming from the field of synthesis how to proceed if characterization of impurities are on focus (**Problem solving with HPLC-MS – a practical view from practitioners**). Oliver Müller (**LC/MS from the perspective of a maintenance engineer**) undertakes a virtual walk across a MS and gives hints how to handle the problem "impurities in LC/MS".

Part IV

Finally, in Part IV (Chapter 8–10, **Report of device manufacturers – article by Agilent, SCIEX, and ThermoScientific**) three manufacturers introduce briefly their newest products and evaluate the future of HPLC-MS coupling.

We think the style and structure of *The HPLC Expert* has proven itself, so those were kept the same in the subsequent book: the book need not to be read linearly. All chapters present self-contained modules – "jumping" between chapters is always possible. That way we try to keep the character of the book as a reference book for LC/MS users. The reader may benefit therefrom.

List of Contributors

Dr. Claudia vom Eyser
Institut für Energie- und
Umwelttechnik e. V.
Bliersheimer Straße 58–60
47229 Duisburg
Germany

Dr. Edmond Fleischer
MicroCombiChem e. K.
Rheingaustraße 190–196
Building E512
65203 Wiesbaden
Germany

Terence Hetzel
Institut für Energie- und
Umwelttechnik e. V.
Bliersheimer Straße 58–60
47229 Duisburg
Germany

Dr. Andreas Hofmann
Novartis
Institutes for BioMedical Research
Novartis Campus
4056 Basel
Switzerland

Dr. Friedrich Mandel
Friedrich-Speidel-Straße 43
76307 Karlsbad
Germany

Dr. Markus M. Martin
Thermo Fischer Scientific
Dornierstraße 4
82110 Germering
Germany

Oliver Müller
Fischer Analytics GmbH
Duhlwiesen 32
55413 Weiler bei Bingen
Germany

Alban Muller
Novartis Institutes for BioMedical
Research
Novartis Campus
4056 Basel
Switzerland

Dr. Christoph Portner
Tauw GmbH
Richard-Löchel-Straße 9
47441 Moers
Germany

Dr. Detlev Schleuder
AB SCIEX Germany GmbH
Landwehrstraße 54
64293 Darmstadt
Germany

Prof. Dr. Oliver J. Schmitz
University of Duisburg–Essen
Faculty of Chemistry
Applied Analytical Chemistry
Campus Essen, S05 T01 B35
Universitätsstr. 5
45141 Essen
Germany

Dr. Terry Sheehan
Director MS Business Development
Agilent Technologiesy
5301 Stevens Creek Blvd, 3U-WI
Santa Clara, CA 95051
USA

Dr. Thorsten Teutenberg
Institut für Energie- und
Umwelttechnik e. V.
Bliersheimer Straße 58–60
47229 Duisburg
Germany

Dr. Jochen Türk
Institut für Energie- und
Umwelttechnik e. V.
Bliersheimer Straße 58–60
47229 Duisburg
Germany

Dr. Steffen Wiese
Institut für Energie- und
Umwelttechnik e. V.
Bliersheimer Straße 58–60
47229 Duisburg
Germany

Part I
Overview, Pitfalls, Hardware-Requirements

1
State of the Art in the LC/MS
O. Schmitz

1.1
Introduction

The dramatically increased demands on the qualitative and quantitative analysis of more complex samples are a huge challenge for modern instrumental analysis. For complex organic samples (e.g., body fluids, natural products or environmental samples), only chromatographic or electrophoretic separations followed by mass spectrometric detection meet these requirements. However, at the moment a tendency can be observed, in which a complex sample preparation and preseparation is replaced by high-resolution mass spectrometer with atmospheric pressure ion sources. However, numerous ion–molecule reactions in the ion source – especially in complex samples due to incomplete separation – are possible because the ionization in typical atmospheric pressure ion sources is nonspecific [1]. Thus, this approach often leads to ion suppression and artifact formation in the ion source, particularly in electrospray ionization (ESI) [2].

Nevertheless, sources such as ASAP (atmospheric pressure solids analysis probe), DART (direct analysis in real time), and DESI (desorption electrospray ionization) can often be successfully used. In ASAP, a hot nitrogen flow from an ESI or APCI (atmospheric pressure chemical ionization) source is used as a source of energy for evaporation and the only change to an APCI source is the installation of an insertion option to place the sample in the hot gas stream within the ion source [3]. This ion source allows a rapid analysis of volatile and semivolatile compounds and, for example, was used to analyze biological tissue [3], polymer additives [3], fungi and cells [4], and steroids, [3, 5]. ASAP has much in common with DART [6] and DESI [7]. The DART ion source produces a gas stream containing long-lived electronically excited atoms that can interact with the sample and, thus, desorption and subsequent ionization of the sample by Penning ionization [8] or proton transfer from protonated water clusters [6] is realized. The DART source is used for the direct analysis of solid and liquid samples. A great advantage of this source is the possibility to analyze compounds on surfaces such as illegal substances on dollar bills or fungicides on wheat [9]. Unlike ASAP and DART, the great advantage of DESI is that the volatility of the analyte is not a

The HPLC-MS Handbook for Practitioners, First Edition. Edited by S. Kromidas.
© 2017 WILEY-VCH Verlag GmbH & Co. KGaA. Published 2017 by WILEY-VCH Verlag GmbH & Co. KGaA.

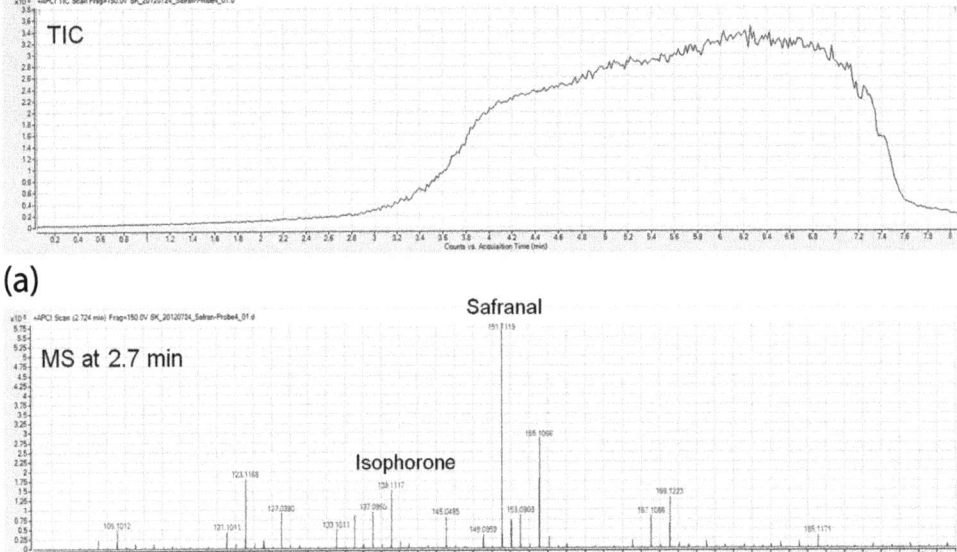

Figure 1.1 Analysis of saffron using direct-inlet probe-APCI with high-resolution QTOF-MS. (a) TIC of the toal analysis. (b) mass spectrum at the time of 2.7 min.

prerequisite for a successful analysis (same as in the classic ESI). DESI is most sensitive for polar and basic compounds and less sensitive for analytes with a low polarity [10]. These useful ion sources have a common drawback. All or almost all substances in the sample are present at the same time in the gas phase during the ionization in the ion source. The analysis of complex samples can therefore lead to ion suppression and artifact formation in the atmospheric pressure ion source due to ion–molecule reactions on the way to the MS inlet. For this reason, some ASAP applications are described in the literature with increasing temperature of the nitrogen gas [5, 11, 12]. DART analyzes with different helium temperatures [13] or with a helium temperature gradient [14] have been described in order to achieve a partial separation of the sample due to the different vapor pressures of the analyte. Related with DART and ASAP, the direct inlet sample APCI (DIP-APCI) from Scientific Instruments Manufacturer GmbH (SIM) was described 2012, which uses a temperature-push rod for direct intake of solid and liquid samples with subsequent chemical ionization at atmospheric pressure [15]. Figure 1.1 shows a DIP-APCI analysis of a saffron sample (solid, spice) without sample preparation with the saffron-specific biomarkers isophorone and safranal. As a detector, an Agilent Technologies 6538 UHD Accurate-Mass Q-TOF was used. The total ion chromatogramm (TIC) of the total analysis and the mass spectrum at the time of 2.7 min are shown in Figure 1.1a,b, respectively. The analysis was started at 40 °C and heated the sample at 1 K/s to a final temperature of 400 °C.

These ion sources may be useful and time saving but for the quantitative and qualitative analysis of complex samples a chromatographic or electrophoretic preseparation makes sense. In addition to the reduction of matrix effects, the comparison of the retention times also allows an analysis of isomers.

1.2
Ionization Methods at Atmospheric Pressure

In the last 10 years, several new ionization methods for atmospheric pressure (AP) mass spectrometers have been developed. Some of these are only available in some working groups. Therefore, only four commercially available ion sources will be presented in detail here.

The most common atmospheric pressure ionization (API) is electrospray ionisation (ESI), followed by APCI and APPI (atmospheric pressure photoionization). A significantly lower significance shows the APLI (atmospheric pressure laser ionization). However, this ion source is well suited for the analysis of aromatic compounds and, for example, the gold standard for PAH (polyaromatic hydrocarbons) analysis. This ranking reflects more or less the chemical properties of the analytes, which are determined with API MS: Most analytes from the pharmaceutical and life sciences are polar or even ionic and, thus, are efficiently ionized by ESI (Figure 1.2). However, there is also a considerable interest in API techniques for efficient ionization of less or nonpolar compounds. For the ionization of such substances ESI is less suitable.

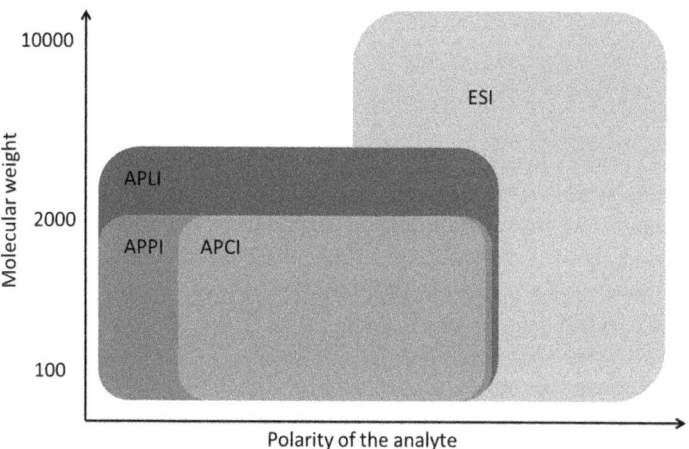

Figure 1.2 Polarity range of analytes for ionization with various atmospheric pressure ionization (API) techniques. Note: The extended mass range of APLI against APPI and APCI results from the ionization of nonpolar aromatic analytes in an electrospray Reproduced with kind permission of O. J. Schmitz, T. Benter, Advances in LC-MS Instrumentation: Atmospheric pressure laser ionization, Journal of Chomatography Libary, Vol 72 (2007), Chapter 6, Pages 89-113.

1.2.1
Overview of API Methods

Ionization methods that operate at atmospheric pressure, such as atmospheric pressure chemical ionization (APCI) and electrospray ionization (ESI), have greatly expanded the scope of mass spectrometry [17–20]. These API techniques allow an easy coupling of chromatographic separation systems, such as liquid chromatography (LC), to a mass spectrometer.

There is a fundamental difference between APCI and ESI ionization mechanism. In APCI, ionization of the analyte takes place in the gas phase after evaporation of the solvent. In ESI, the ionization takes place already in the liquid phase. In the ESI process, protonated or deprotonated molecular ions are usually formed from highly polar analytes. Fragmentation is rarely observed. However, for the ionization of less polar substances, APCI is preferably used. APCI is based on the reaction of analytes with primary ions, which are generated by corona discharge. But the ionization of nonpolar analytes is very low with both techniques.

For these classes of substances other methods have been developed, such as the coupling of ESI with an electrochemical cell [21–32], the "coordination ionspray" [32–47] or the "dissociative electron-capture ionization" [38–42]. The atmospheric pressure photoionization (APPI) or the dopant-assisted (DA) APPI presented by Syage et al. [43, 44] and Robb et al. [45, 46], respectively, are relatively new methods for photoionization (PI) of nonpolar substances by means of vacuum ultraviolet (VUV) radiation. Both techniques are based on photoionization, which is also used in ion mobility mass spectrometry [47–50] and in the photo ionization detector (PID) [51–53].

1.2.2
ESI

In the past, one of the main problems of mass spectrometric analysis of proteins or other macromolecules was that their mass was outside the mass range of most mass spectrometers. For the analysis of larger molecules, such as proteins a hydrolysis and the analysis of the resulting peptide mixture had to be carried out. With ESI it is now possible to ionize large biomolecules without prior hydrolysis and analyze them by MS.

Based on previous works from Zeleny [54], Wilson and Taylor [55, 56], Dole et al. produced high molecular weight polystyrene ions in the gas phase from a benzene/acetone mixture of the polymer by electrospray [57]. This ionization method was finally established through the work of Fenn in 1984 [58], who was awarded the Nobel Prize for Chemistry in 2002.

In order to describe the whole process of ion formation in ESI, a subdivision of processes into three sections makes sense:

- Formation of charged droplets
- Reduction of the droplet
- Formation of gaseous ions.

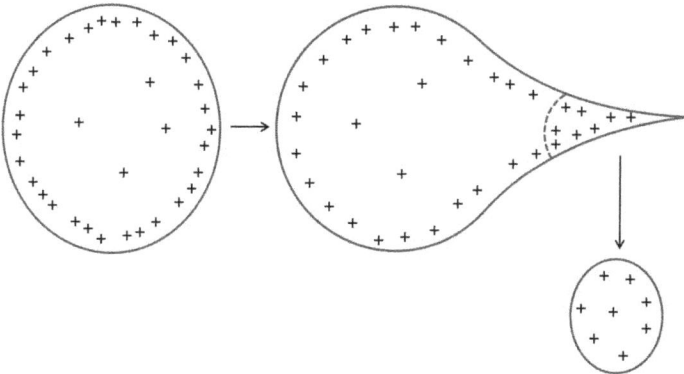

Figure 1.3 Reduction of the droplet size.

To generate positive ions a voltage of 2–3 kV between the narrow capillary tip (10^{-4} m outer diameter) and the MS input (counter electrode) is applied. In the exiting eluate from the capillary, charge separation occurs. Cations are enriched at the surface of the liquid and moved to the counter electrode. Anions migrate to the positively charged capillary, where they are discharged or oxidized. The accumulation of positive charge on the liquid surface is the cause of the formation of a liquid coned, as the cations are drawn to the negative pole, the cathode. This so-called Taylor cone resulted from the electric field and the surface tension of the solution. With certain distance from the capillary, there is a growing destabilization and a stable spray of drops with an excess of positive charges will emitted.

The size of the droplets formed is dependent on the

- Flow rate of the mobile phase and the auxiliary gas
- Surface tension
- Viscosity
- Applied voltage and
- Concentration of the electrolyte.

These drops lose solvent molecules by evaporation and at the Raleigh limit (electrostatic repulsion of the surface charges > surface tension) much smaller droplets (so-called microdroplets) are emitted (Figure 1.3). This occurs due to elastic surface vibrations of the drops which lead to formation of Taylor cone-like structures.

At the end of such protuberances small droplets are formed, which have a significantly smaller mass/charge ratio than the "mother drop". Because of the unequal decomposition the ratio of surface charge to the number of paired ions in the droplet increases dramatically per cycle of droplet formation and evaporation up to the Raleigh limit in comparison with the "mother drops". Thus, only highly charged microdroplets are responsible for the successful formation of ions.

For the ESI process, the formation of multiply charged ions for large analyte molecules is characteristic. Therefore, a series of ion signals for, for example, peptides and proteins can be observed, which differ to each other by one charge (usually an addition of a proton in positive mode or subtraction of a proton in negative mode).

For the formation of the gaseous analyte two mechanisms are discussed at the moment. The charged residue mechanism (CRM) proposed by Cole [59] and Kebarle and Peschke [60] and the ion evaporation mechanism (IEM) postulated by Thomason and Iribarne [61]. In CRM, the droplets are reduced as long as only one analyte in the microdroplets is present, then one or more charges are added to the analyte. In IEM, the droplets are reduced to a so-called critical radius ($r < 10$ nm) and then charged analyte ions are emitted from these drops [62]. It is essential for the process that enough charge carriers are provided in the eluate. This can be realized by the addition of, for example, ammionium formiate to the eluent or eluate. Without this addition, ESI is also possible with an eluate of acetonitrile/water (but not with MeOH/water), but a more stable and more reproducible electrospray with a higher ion yield is only formed by adding charge carriers before or after HPLC separation.

1.2.3
APCI

This ionization method was developed by Horning in 1974 [63]. The eluate is introduced through an evaporator (400–600 °C) into the ion source. Despite the high temperature of the evaporator, a decomposition of the sample is only rarely observed because the energy is used for the evaporation of the solvent, and the sample is not normally heated above 80 to 100 °C [64]. In the exit area of the gas flow (eluate and analyte), a metal needle (corona) is mounted to which a high voltage is applied. When the solvent molecules reach the field of high voltage, a reaction plasma is formed on the principle of chemical ionization. If the energy difference between the analyte and reactant ions is large enough, the analytes are ionized, for example, by proton transfer or adduct formation in the gas phase.

For the emission of electrons in APCI a corona discharge is used instead the filament in GC/MS (CI) because of the rapid fusing of the filament at atmospheric pressure. In APCI, with nitrogen as sheath and nebulizer gas and atmospheric water vapor (is also available in 5.0 nitrogen sufficient quantity), $N_2^{+\bullet}$ and $N_4^{+\bullet}$ ions are primarily formed by electron ionization. These ions collide with the vaporized solvent molecules and form secondary reactant gas ions, such as H_3O^+ and $(H_2O)nH^+$ (Figure 1.4).

$$N_2 + e^- \rightarrow N_2^{+\bullet} + 2e^-$$
$$N_2^{+\bullet} + 2N_2 \rightarrow N_4^{+\bullet} + N_2$$
$$N_4^{+\bullet} + H_2O \rightarrow H_2O^{+\bullet} + 2N_2$$
$$H_2O^{+\bullet} + H_2O \rightarrow H_3O^+ + OH^-$$

$$H_3O^+ + n\, H_2O \rightarrow H_3O^+ \cdot (H_2O)_n$$

Figure 1.4 Reaction mechanism in APCI.

The most common secondary cluster ion is $(H_2O)_2H^+$, together with significant amounts of $(H_2O)_3H^+$ and H_3O^+. These charged water clusters collide with the analyte molecules, resulting in the formation of analyte ions:

$$H_3O + M \rightarrow [M + H]^+ + H_2O \tag{1.1}$$

The high collision frequency results in a high ionization efficiency of the analytes and adduct ions with little fragmentation. In the negative mode, the electrons that are emitted during the corona discharge form together with large amounts of N_2 and the presence of water molecules OH^- ions. Due to the fact that the gas phase acidity of H_2O is very low, OH^- ions in the gas phase form by proton transfer reaction with the analyte H_2O and $[MH]^-$ (M = analyte) [64]. The problem with APCI is the simultaneous formation of different adduct ions. Depending on eluent composition and matrix components, it is possible that Na^+ and NH_4^+ adducts are formed in addition to protonated analyte molecules, making the data evaluation more difficult.

1.2.4
APPI

APPI is suitable for the ionization of nonpolar analytes, in which the photoionization of molecule M leads to the formation of a radical cation $M^{\bullet+}$. If the ionization potentials (IPs) of all other matrix elements are greater than the photon energy, then the ionization process is specific for the analyte. However, in the APPI different processes can very strongly influence the detection of $M^{\bullet+}$:

1. In the presence of solvent molecules and/or other existing components in large excess, ion–molecule reactions can proceed.
2. VUV photons are efficiently absorbed from the gas phase matrix.

Thus, for example, in the presence of acetonitrile (a commoonly used mobile phase in HPLC) mainly $[M + H^+]$ is formed even though the IP of acetonitrile is more than 2.2 eV higher as the photon energy [65]. In general, in the case of polar compounds, which are dissolved in CH_3CN/H_2O, the formation of $[M + H]^+$ is usually observed, while nonpolar compounds such as naphthalene, usually form $M^{\bullet+}$ [66]. A detailed mechanism for the formation of $[M + H]^+$ was proposed by Klee et al. [67]. In APPI, the ion yield is reduced due to the limited VUV photon flux, and the interactions with solvent molecules. Therefore, the dopant-assisted atmospheric pressure photo ionization (DA-APPI) was introduced as a new ionization method from Bruins et al. [66].

The total number of ions which are formed by the VUV radiation is significantly increased by the addition of a directly ionizing component (dopant). If the dopant is selected such that the resulting dopant ions have a relatively high recombination energy or low proton affinity, then the dopant ion can ionize the analytes by charge exchange or proton transfer. In addition to acetone and toluene, anisole was also found to be a very effective dopant in APPI [68]. By adding a dopant the sensitivity can be increased, but the possible adduct formations often lead to significantly more complicated APPI mass spectra [45, 66, 68]. Recent studies suggest that the direct proton transfer from the initially formed dopant ions plays only a very minor role, and the ionization process is dominated by a very complex, thermodynamically controlled cluster chemistry.

1.2.5
APLI

Atmospheric pressure laser ionization (APLI) was developed in 2005 [69]. It is a soft ionization method with easy-to-interpret spectra for nonpolar aromatic substances and only minor tendency for fragmentation of the analytes. APLI is based on the resonance-enhanced multiphoton ionization (REMPI), however, at atmospheric pressure.

The REMPI method allows the sensitive and selective ionization of numerous compounds. Here, for example, the following approach is used:

$$M + mh\nu \rightarrow M^* \tag{1.2}$$

$$M^* + nh\nu \rightarrow M^{\bullet +} + e^- \tag{1.3}$$

Reactions (1.1) and (1.2) represent a classical $(m + n)$ resonance-enhanced multiphoton ionization (REMPI) process, which $n = m = 1$ is often very beneficial used for the ionization of polyaromatic hydrocarbons (PAH). Because the absorption bands of PAHs are relatively broad at room temperature and PAHs have high molecular absorption coefficient in the near ultraviolet and a relatively long lifetime of the S1 and S2 states, a fixed frequency laser, for example the 248 nm line of a KrF excimer laser, can be used. Under these conditions, an almost selective ionization of aromatic hydrocarbons can be achieved.

A great advantage of APLI in comparison to APPI is that neither oxygen nor nitrogen and the solvents typically used in the HPLC (for example, water, methanol, acetonitrile) have appreciable absorption cross sections in the used wavelength range. An attenuation of the photon density within the ion source, that is, a significant coupling of electronic energy into the matrix, as observed in the APPI, does not take place in APLI. The APLI is very sensitive in the determination of PAHs and, therefore, represents a valuable alternative to APCI and APPI, but APLI is not only restricted to the analysis of such simple aromatic compounds. More complex oligomeric or polymeric structures, and organometallic compounds can also be analyzed [70]. It is also possible to analyze nonaromatic compounds after derivatization of their functional group with so-called ionization markers, in analogy to fluorescence derivatization [71]. With this technique you can benefit from the selectivity of the ionization (only aromatic systems) and the outstanding sensitivity of the method. In addition, a parallel ionization of sample components with ESI or APCI together with APLI was realized [72–74] to analyze polar (ESI) or nonaromatic medium polar (APCI) together with aromatic (APLI) compounds.

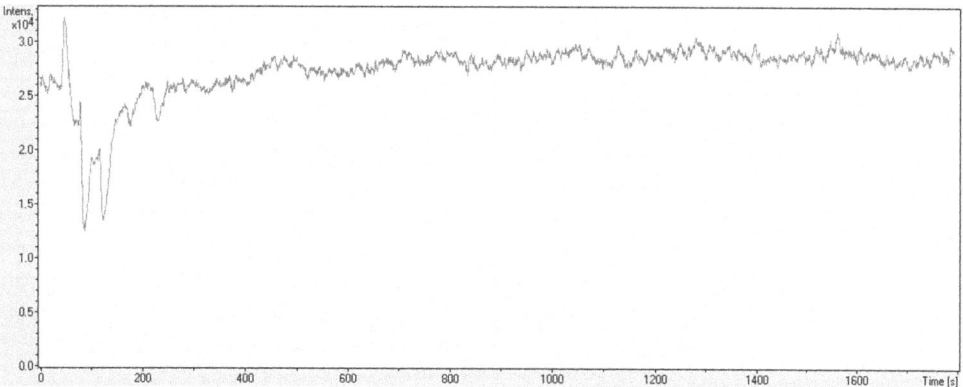

Figure 1.5 Ion suppression in APCI-MS of PAH in urine.

1.2.6
Determination of Ion Suppression

In many mass spectrometric analyses of complex samples, ion suppression leads to a more difficult quantitative determination and time-consuming sample preparation is often required. It should therefore be studied more in advance whether there is a signal-reducing influence of the matrix.

For the investigation of ion suppression, the sample solution (without analyte) is injected in the HPLC and a solution with the analyte (stable-isotopic labeled analyte, if no sample solution without analyte is available) is mixed behind the separation column via a T-piece to the eluate and the mass trace of the analyte (or stable-isotopic labeled analyte) is analyzed during the total analysis time. After the column, the separated matrix ingredients are mixed with the analyte in the T-piece and are transported into the ion source. The change in intensity of the analyte mass trace before and after the injection of the matrix provides information about a possibly occurring ion suppression.

Figure 1.5 shows the determination of ion suppression of a PAH analysis in urine with APCI-QTOF. During the analysis time between 80 and 400 s, the mass trace is considerably diminished and reached the normal level after about 450 s. This means that disturbing matrix components in the urine left the column between 80 and 400 s, which leads to ion suppression.

1.2.7
Best Ionization for Each Question

On the basis of Figure 1.2, the method which allows the most effective ionization for the analyte of interest can roughly be estimated. Depending on the polarity of the analyte, the ionization should be done with ESI (polar analytes), APCI (moderately polar analytes), APPI (nonpolar analytes), or with APLI (aromatics). However, the matrix plays an important role in making this decision. For complex samples, ion suppression with electrospray ionization is more likely and more pronounced than for the other ionization methods discussed here. The ion beam line plays also an important role in the inlet region of the mass spectrometer. ESI ion sources with a Z-spray inlet show often less ion suppression than normal ESI ion sources. Also, the eluate flow must be adapted to the ion source. For example, slightly higher fluxes than with ESI sources can often be used in APCI sources. Although equipment manufacturers promise other flow rates, it is useful to operate ESI sources with fluxes below 300 µl/min and APCI, APPI, and APLI sources with fluxes below 500 µl/min with regard to spray stability, reproducibility, and ion suppression. Of course, based on the application even larger flows can be used, but often problems such as ion suppression or spray instability are observed.

1.3
Mass Analyzer

The most frequent mass spectrometers, which are routinely coupled to the LC:

- Quadrupole
- TripleQuad
- IonTrap
- TOF
- Orbitrap.

With regard to sensitivity and ratio of price and performance (including maintenance), a quadrupole MS is a very good purchase. With single ion mode (SIM), a very good sensitivity can be achieved and a fast quadrupole (from about 25–50 Hz) allows coupling with a fast UHPLC separation.

Based on quadrupole MS, a further development represents the triple quadrupole mass spectrometer, which play an increasingly important role, especially in the target analysis in complex samples. The sample preparation is minimized, a preliminary separation is often omitted and the potentials of the first and third quadrupole are adjusted so that only a certain mass is allowed to pass these quadrupoles. In the first quadrupole, the ion of the target analyte and in the third quadrupole a characteristic ion fragment, which is induced by collisions with argon in the second quadrupole (actually a fragmentation cell) is passed through. Due to the analysis of the fragment ion, the chemical noise (matrix) is greatly reduced and the triple quadrupole mass spectrometers are one of the most sensitive and selective mass spectrometers. Detection limits in zeptomoles area (amount of substance on the separation column) have been realized for some analytes.

Similar to a quadrupole, an ion trap is constructed. However, the ions are collected in the trap, and then, either a mass scan or single to multiple fragmentation of the target analyte can be performed. Modern ion trap MS systems are characterized by a very good linearity and sensitivity and a fast data acquisition (e.g., 20 Hz) and, thus, can even be coupled with UHPLC. They are particularly suitable for structure determination of biomolecules (e.g., carbohydrates, peptides).

For more as 20 years, the use of time-of-flight (TOF) mass spectrometers is increasing, which is related to the orthogonal ion beam guiding in the device. The orthogonal ion beam has made it possible to couple even continuous ion sources, such as ESI and APCI, without loss of resolution to a TOF-MS. Recently, the resolution was steadily improved through the introduction of repeller electrodes, ion funnels, and more powerful electronics etc., so that now several manufacturers offer TOF-MS systems with resolutions up to 50 000 while realizing data acquisition rates of 20 Hz or more. Thus, these devices are ideally suitable for the coupling of fast separation techniques such as UHPLC and can also provide assistance in the identification of unknown sample components due to the high resolution and mass accuracy (< 1 ppm).

One of the latest mass analyzer is the LTQ Orbitrap mass spectrometer (LTQ = linear trap quadrupole). In this, the commercial LTQ is coupled with an ion trap, developed by Makarov [75, 76]. Due to the resolving power (between 70 000 and 800 000) and the high mass accuracy (1–3 ppm), Orbitrap mass analyzers for example be used for identification of peptides in protein analysis or for metabolomic studies. In addition, the selectivity of MS/MS experiments can be greatly improved. However, the coupling is not useful with UHPLC for rapid chromatographic preseparation, as the data acquisition rate is too low for a reproducible integration of the narrow signals produced with UHPLC.

In addition to some other mass spectrometers, FT-ICRMS devices are also used. The latter, in addition to very high acquisition and operating costs (e.g., helium), has the disadvantage of low data acquisition rate (same problem as with the Orbitrap), so the coupling with a fast analysis, such as UHPLC cannot be realized. However, they are unbeaten in resolution (> 800 000) and an extremely useful tool in metabolomic research.

1.4
Future Developments

The trend in mass spectrometry is currently clearly toward higher resolution and faster data acquisition.

Probably in future resolution of about 100 000 and data rates of 20–40 Hz can be achieved with TOF-MS. With Orbitrap-MS, it is assumed that resolutions of more than 500 000 will be possible by more precise production of the cell and electronic. This could then by shortening the scanning speed, which is accompanied by a loss of resolution, allow a fast preseparation with UHPLC.

In the area of nontarget analysis, the combination of ion mobility spectrometry (IMS) with a high resolution QTOF-MS presents a powerful analysis platform. Two commercial systems with different varieties of ion mobility methods – the drift time ion mobility spectrometry (DTIM) from Agilent 6560 and the traveling wave ion mobility spectrometry (TWIMS) from Waters (Vion IMS QTOF) – are currently available. Due to the structure-dependent drift time in the drift tube of the IMS, isobaric substances can be separated from each other. Figure 1.6 shows

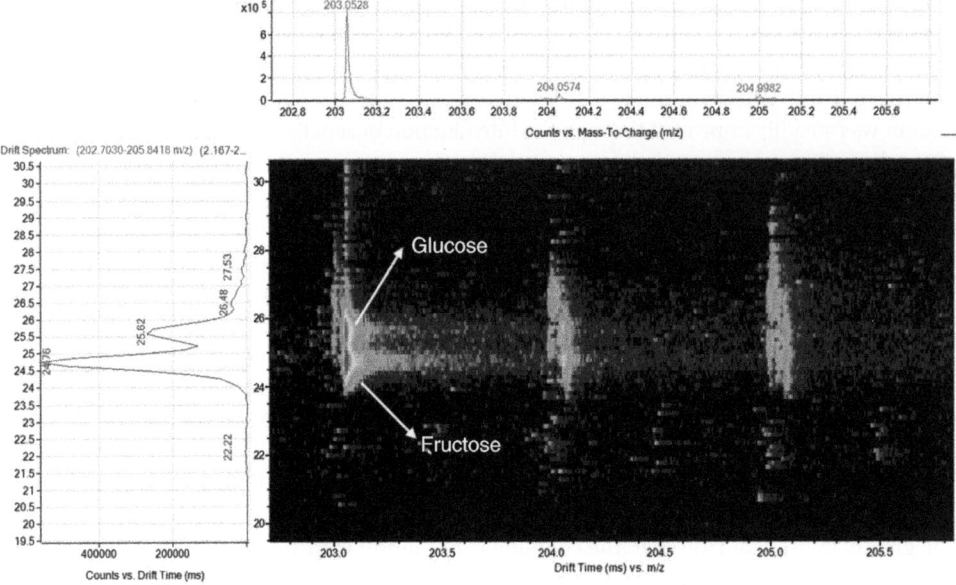

Figure 1.6 Analysis of a mixture of glucose and fructose with IM-qTOF-MS.

the separation of two isobaric substances, glucose and fructose, in the IM-QTOF-MS system (Agilent 6560) by their different drift time (in ms) in the 80 cm long drift tube of the system. Particularly noteworthy is that the collision cross section (CCS) of substances can directly (the Agilent system) or indirectly (through comparison with a standard, the Waters system) be determined with the help of the drift time. With a database of CCS values and precise mass, a fast and reliable identification of the signals can then be carried out for a nontarget analysis.

Another focus in future developments will be the optimization of ion sources with respect to ion generation and ion transport at different flows which are used in nano- and micro-HPLC, LC×LC and supercritical fluid chromatography (SFC) to increase the sensitivity.

1.5
What Should You Look for When Buying a Mass Spectrometer?

In addition to the available budget, in my opinion the following points playing a central role for a buying decision:

- Should a target analysis or comprehensive analysis of the sample be carried out
- Needed sensitivity
- Software
- Sample throughput
- MS analysis with or without preseparation process.

If only target analyzes are planned (e.g., analysis of known impurities in a product or pesticide analysis), a quadrupole or triple quadrupole-MS would be the best choice. With these devices a very sensitive analysis will be guaranteed and also a quick preseparation (e.g., UHPLC) is now possible for many devices.

If nontarget analysis should be realized, high-resolution mass spectrometer like QTOF or Orbitrap would facilitate the analysis considerably. Due to the additional separation dimension and the determination of CCS values, the new MS systems with an upstream ion mobility spectrometer are certainly an interesting alternative. Even if a high sample throughput is still necessary, the QTOF would have precedence over the slow Orbitrap in high resolution mode. However, regarding the resolution, Orbitrap is the more powerful system compared with QTOF. The sensitivity of qTOF is about a factor 10 lower than that of a triplequad, but detection limits in the lower ppb range are quite possible.

Perhaps, due to a high number of samples, no preseparation will be done. But then it should be ensured that suitable so-called ambient desorption ionization techniques, for example, DESI, DART, ASAP, DIP-APCI, can be coupled to the MS.

Finally, there are large differences in the respective MS software. Here, the user should determine the strengths and weaknesses of the various software systems.

In addition to the price of the system, the operating costs should also be considered. Besides a high nitrogen consumption, the mass spectrometer should be serviced annually. Depending on the effort and manufacturer, maintenance alone leads to an annual cost of 5000–20 000 euros.

References

1 Matuszewski, B.K., Constanzer, M.L., and Chavez-Eng, C.M. (2003) *Anal. Chem.*, **75**, 3019–3030.
2 Annesley, T.M. (2003) *Clin. Chem.*, **49**, 1041–1044.
3 McEwen, C.N., Mckay, R.G., and Larsen, B.S. (2005) *Anal. Chem.*, **77**, 7826–7831.
4 McEwen, C. and Gutteridge, S. (2007) *J. Am. Soc. Mass. Spectrom.*, **18**, 1274–1278.
5 Ray, A.D., Hammond, J., and Major, H. (2010) *Eur. J. Mass. Spectrom.*, **16**, 169–174.
6 Cody, R.B., Laramee, J.A., and Durst, H.D. (2005) *Anal. Chem.*, **77**, 2297–2302.
7 Takats, Z., Wiseman, J.M., Gologan, B., and Cooks, R.G. (2004) *Science*, **306**, 471–473.
8 Laramee, J.A. and Cody, R.B., (2007) in *The Encyclopedia of Mass Spectrometry*, (eds M.L. Gross and R.M. Caprioli), vol. 6, Elsevier.
9 Schurek, J., Vaclavik, L., Hooijerink, H., Lacina, O., Poustka, J., Sharman, M., Caldow, M., Nielen, M.W.F., and Hajslova, J. (2008) *Anal. Chem.*, **80**, 9567–9575.
10 Lloyd, J.A., Harron, A.F., and Mcewen, C.N. (2009) *Anal. Chem.*, **81**, 9158–9162.
11 Ahmed, A., Cho, Y.J., No, M.H., Koh, J., Tomczyk, N., Giles, K., Yoo, J.S., and Kim, S. (2011) *Anal. Chem.*, **83**, 77–83.
12 Pan, H.F. and Lundin, G. (2011) *Eur. J. Mass. Spectrom.*, **17**, 217–225.
13 Maleknia, S.D., Vail, T.M., Cody, R.B., Sparkman, D.O., Bell, T.L., and Adams, M.A. (2009) *Rapid Commun. Mass Spectrom.*, **23**, 2241–2246.

14 Edison, S.E., Lin, L.A., Gamble, B.M., Wong, J., and Zhang, K. (2011) *Rapid Commun. Mass Spectrom.*, **25**, 127–139.

15 Krieger, S., von Trotha, A., Leung, K.S.-Y., and Schmitz, O.J. (2013) Analytical and bioanalytical chemistry. *Anal. Bioanal. Chem.*, **405**, 1373, doi:10.1007/s00216-012-6531-4.

16 Schmitz, O.J. and Benter, T. (2007) in *Advances in LC-MS Instrumentation: Atmospheric pressure laser ionization*, vol. 72, Chapter 6, Journal of Chromatography Library, Elsevier, Amsterdam, pp. 89–113.

17 Cole, R.B. (ed.) (1997) *Electrospray Ionization Mass Spectrometry*, John Wiley & Sons, Inc., New York.

18 Cech, N.B. and Enke, C.G. (2001) *Mass Spectrom. Rev.*, **20**, 362–387.

19 Kebarle, P. (2000) *J. Mass. Spectrom.*, **35**, 804–817.

20 Niessen, W.M.A. (ed.) (1999) *Liquid Chromatography – Mass Spectrometry*, Marcel Dekker, Inc., New York.

21 Van Berkel, G.J., McLuckey, S.A., and Glish, G.L. (1991) *Anal. Chem.*, **63**, 2064–2068.

22 Van Berkel, G.J., McLuckey, S.A., and Glish, G.L. (1992) *Anal. Chem.*, **64**, 1586–1593.

23 Van Berkel, G.J. and Asano, K.G. (1994) *Anal. Chem.*, **66**, 2096–2102.

24 Van Berkel, G.J. and Zhou, F. (1995) *Anal. Chem.*, **67**, 2916–2923.

25 Van Berkel, G.J. and Zhou, F. (1995) *Anal. Chem.*, **67**, 3958–3964.

26 Van Berkel, G.J., Quirke, J.M.E., Tigani, R.A., Dilley, A.S., and Covey, T.R. (1998) *Anal. Chem.*, **70**, 1544–1554.

27 Van Berkel, G.J., Quirke, J.M.E., and Adams, C.L. (2000) *Rapid Commun. Mass Spectrom.*, **14**, 849–858.

28 Williams, D. and Young, M.K. (2000) *Rapid Commun. Mass Spectrom.*, **14**, 2083–2091.

29 Quirke, J.M.E., Hsz, Y.-L., and Van Berkel, G.J. (2000) *Nat. Prod.*, **63**, 230–237.

30 Williams, D., Chen, S., and Young, M.K. (2001) *Rapid Commun. Mass Spectrom.*, **15**, 182–186.

31 Quirke, J.M.E. and Van Berkel, G.J. (2001) *J. Mass Spectrom.*, **36**, 179–187.

32 Kauppila, T.J., Kostiainen, R., and Bruins, A.P. (2004) *Rapid Commun. Mass Spectrom.*, **18**, 808–815.

33 Rentel, C., Strohschein, S., Albert, K., and Bayer, E. (1998) *Anal. Chem.*, **70**, 4394–4400.

34 Bayer, E., Gfrörer, P., and Rentel, C. (1999) *Angew. Chem. Int. Ed.*, **38**, 992–995.

35 Takino, M., Daishima, S., Yamaguchi, K., and Nakahara, T. (2001) *J. Chromatogr. A*, **928**, 53–61.

36 Roussis, S.G. and Proulx, R. (2002) *Anal. Chem.*, **74**, 1408–1414.

37 Marwah, A., Marwah, P., and Lardy, H. (2002) *J. Chromatogr. A*, **964**, 137–151.

38 Singh, G., Gutierrez, A., Xu, K., and Blair, I.A. (2000) *Anal. Chem.*, **72**, 3007–3013.

39 Higashi, T., Takido, N., Yamauchi, A., and Shimada, K. (2002) *Anal. Sci.*, **18**, 1301–1307.

40 Higashi, T., Takido, N., and Shimada, K. (2003) *Analyst*, **128**, 130–133.

41 Hayen, H., Jachmann, N., Vogel, M., and Karst, U. (2002) *Analyst*, **127**, 1027–1030.

42 Zwiener, C. and Frimmel, F.H. (2004) *Anal. Bioanal. Chem.*, **378**, 851–861.

43 Syage, J.A. and Evans, M.D. (2001) *Spectroscopy*, **16**, 15–21.

44 Syage, J.A., Hanold, K.A., Evans, M.D., and Liu, Y. (2001) Atmospheric pressure photoionizer for mass spectrometry, Patent no. WO0197252.

45 Robb, D.B., Covey, T.R., and Bruins, A.P. (2000) *Anal. Chem.*, **72**, 3653–3659.

46 Robb, D.B. and Bruins, A.P. (2001) Atmospheric pressure photoionization (APPI): A new ionization method for liquid chromatography – mass spectrometry, Patent No. WO0133605.

47 Baim, M.A., Eartherton, R.I., and Hill, H.H. Jr. (1983) *Anal. Chem.*, **55**, 1761–1766.

48 Leasure, C.S., Fleischer, M.E., Anderson, G.K., and Eiceman, G.A. (1986) *Anal. Chem.*, **58**, 2142–2147.

49 Spangler, G.E., Roehl, J.E., Patel, G.B., and Dorman, A. (1994) US Patent no. 5,338,931.

50 Kauppila, T.J., Kuuranne, T., Meurer, E.C., Eberlin, M.N., Kotiaho, T., and

Kostiainen, R. (2002) *Anal. Chem.*, **74**, 5470–5479.

51 Discroll, J.N. (1976) *Am. Lab.*, **8**, 71–75.

52 Discroll, J.N. (1977) *J. Chromatogr.*, **134**, 49–55.

53 Locke, D.C., Dhingra, B.S., and Baker, A.D. (1982) *Anal. Chem.*, **54**, 447–450.

54 Zeleny, J. (1917) *Phys. Rev.*, **10**, 1–6.

55 Wilson, C.T.R. and Taylor, G. (1925) *Proc. Cambridge Philos. Soc.*, **22**, 728–730.

56 Taylor, G. (1964) *Proc. R. Soc. Lond. Ser. A.*, **280**, 383–397.

57 Dole, M., Mack, L.L., Hines, R.L., Mobley, R.C., Ferguson, L.D., and Alice, M.B. (1968) *J. Chem. Phys.*, **49**, 2240–2249.

58 Yamashita, M. and Fenn, J.B. (1984) *J. Phys. Chem.*, **88**, 4451–4459.

59 Cole, R.B. (2000) *J. Mass Spectrom.*, **35**, 763–772.

60 Kebarle, P. and Peschke, M. (1994) *Anal. Chem.*, **66**, 712–718.

61 Thomson, B.A. and Iribarne, J.V. (1979) *J. Chem. Phys.*, **71**, 4451–4463.

62 Molin, L. and Traldi, P. (2007) in *Advances in LC-MS Instrumentation, Basic Aspects of Electrospray Ionization*, (ed. A. Cappiello), vol. 72, Chapter 1, Journal of Chromatography Library, Elsevier Science, pp. 1–9.

63 Carrol, D.I., Dzidic, I., Stillwell, R.N., Horning, M.G., and Horning, E.C. (1974) *Anal. Chem.*, **46**, 706–710.

64 Moini, M. (2007) in *The Encyclopedia of Mass Spectrometry, Atmospheric Pressure Chemical Ionization: Principles, Instrumentation, and Applications* (eds M.L. Gross and R.M. Caprioli), vol. 6 Elsevier, pp. 344–354

65 Lias, S.G. (2003) Ionization energy evaluation, in *NIST Chemistry WebBook, NIST Standard Reference Database Number 69*, (eds P.J. Linstrom and W.G. Mallard), March 2003, National Institute of Standards and Technology, Gaithersburg MD, 20899 (http://webbook.nist.gov (accessed February 2017).

66 Raffaelli, A. and Saba, A. (2003) *Mass Spectrom. Rev.*, **22**, 318–331.

67 Syage, J.A. (2004) *J. Am. Soc. Mass Spectrom.*, **15**, 1521–1533.

68 Kauppila, T.J., Kotiaho, T., Kostiainen, R., and Bruins, A.P. (2004) *J. Am. Soc. Mass Spectrom.*, **15**, 203–211.

69 Klee, S., Albrecht, S., Derpmann, V., Kersten, H., and Benter, T. (2013) *Anal. Bioanal. Chem.*, **405**, 6933–6951.

70 Constapel, M., Schellenträger, M., Schmitz, O.J., Gäb, S., Brockmann, K.-J., Giese, R., and Benter, T. (2005) *Rapid Commun. Mass Spectrom.*, **19**, 326–336.

71 Tian, N., Thiessen, A., Schiewek, R., Schmitz, O.J., Hertel, D., Meerholz, K., and Holder, E. (2009) *J. Organ. Chem.*, **74**, 2718–2725.

72 Schiewek, R., Mönnikes, R., Wulf, V., Gäb, S., Brockmann, K.J., Benter, T., and Schmitz, O.J. (2008) *Angew. Chem. Int. Ed.*, **47**, 9989–9992

73 Schiewek, R., Mönnikes, R., Wulf, V., Gäb, S., Brockmann, K.J., Benter, T., and Schmitz, O.J. (2008) *Angew. Chem.*, **120**, 10138–10142.

74 Deibel, E., Klink, D., and Schmitz, O.J. (2015) *Anal. Bioanal. Chem.*, **407**, 7425–7434.

75 Schiewek, R., Schellenträger, M., Mönnikes, R., Lorenz, M., Giese, R., Brockmann, K.-J., Gäb, S., Benter, T., and Schmitz, O.J. (2007) *Anal. Chem.*, **79**, 4135–4140.

76 Schiewek, R., Lorenz, M., Brockmann, K.J., Benter, T., Gäb, S., and Schmitz, O.J. (2008) *Anal. Bioanal. Chem.*, **392**, 87–96.

77 Makarov A. (2000) *Anal. Chem.*, **72**, 1156–1162.

78 Perry R.H. et al. (2008) *Mass Spectrom. Rev.*, **27**, 661–699.

2
Technical Aspects and Pitfalls of LC/MS Hyphenation
M.M. Martin

For almost two decades, the coupling of liquid chromatography (LC) and mass spectrometry (MS) has left the stage of breadboard lab designs and is commercialized with manifold off-the-shelf products. Frankly, the first systems on the market required a strong expertise and highly skilled users and, thus, were exclusively applied in highly specialized research laboratories; however, due to intensive research and development work, the robustness and ease of use of LC/MS systems have improved so much over the years that LC/MS techniques are meanwhile established even in many routine applications. Considering how different the two worlds of a separation in the liquid phase via LC and in the gas phase via MS are, this is truly a remarkable fact. Both liquid chromatographs and mass spectrometers have meanwhile achieved a high degree of technical perfection that allows even the less experienced users to obtain reliable results in a fairly short learning time; nevertheless, the list of potential error sources in the LC/MS hyphenation is still long these days. It starts with the selection of unsuitable instrumentation and does not yet end with the wrong interpretation of experimental results. Some errors are specific for instruments, methods, or applications, resulting for instance from specific device features, sample preparation procedures, or critical analyte properties; their individual discussion is beyond the scope of this section. Other aspects are more of a general or fundamental nature. Starting with general considerations on instrumental requirements for a successful LC/MS analysis, this chapter discusses ways for LC/MS method development and adaptation, before it closes with a comprehensive analysis of what can go wrong so that the LC/MS result does not look like the user expects it.

The HPLC-MS Handbook for Practitioners, First Edition. Edited by S. Kromidas.
© 2017 WILEY-VCH Verlag GmbH & Co. KGaA. Published 2017 by WILEY-VCH Verlag GmbH & Co. KGaA.

2.1
Instrumental Requirements for LC/MS Analysis – Configuring the Right System for Your Analytical Challenge

2.1.1
(U)HPLC and Mass Spectrometry – Not Just a Mere Front-End

As with any challenge, the first decision to be made is about the best fitted equipment, so it is always the question of the analytical goal and which LC system is the most appropriate one to achieve success. Do you look into a high-resolution analysis that needs to separate very complex samples as comprehensively as possible into baseline-separated compound zones? Or is a fast, high-throughput capable separation with shortest cycle time that what you need? For a maximum in high resolution, clearly UHPLC systems providing 11 000–22 000 psi/800–1500 bar are of most advantage. Instruments of that performance range have enough reserves to maximize chromatographic resolution or peak capacity by applying very long columns or column chains of 75–100 cm, sub-2-µm particles and long, flat gradients for best mass spectrometric detectability. High-speed, high-throughput analyses, in contrast, are typically run on rather short LC columns of 20–50 mm in length. For the screening of moderately complex sample mixtures like, for example, in the process control of synthesis reactions, stationary phases based on 3-µm particles still offer sufficient resolution, which means that these columns will not generate a high backpressure even at flow rates of more than 1 mL/min, which is already a bit exotic for LC/MS applications anyway. As a consequence, this type of analysis can already be realized on LC systems with moderate UHPLC capabilities in the 9000 psi/600 bar range. Classical HPLC instruments, however, are barely recommended for LC/MS applications, not so much because of the pressure rating, but due to their fairly large gradient delay volume (GDV) and extra-column volume (ECV). For true high-resolution analyses, their performance is too low, while high-throughput applications will suffer from both long gradient delay times and excessive band broadening. Consequently, UHPLC systems have prevailed in many LC/MS applications, particularly in research and development. It is primarily LC/MS analysis which profits substantially from the extended capabilities of the UHPLC technology: a high chromatographic resolution minimizes coelution of analytes and reduces competing ionization and ion suppression in the MS ion source, thus, enhancing sensitivity in the mass spectrometric detection. High-speed analyses with very short run times, in return, improve the workload balance and reduce idle times on a mass spectrometer thus leading to a faster return on investment. However, a thorough look must be given to the configuration and optimization of these systems to translate the high separation performance of the UHPLC column into a high-quality LC/MS chromatogram.

2.1.2
UHPLC System Optimization – Gradient Delay and Extra-column Volumes

From a technical perspective, UHPLC systems are qualified by much more than sheer pressure, although this is probably the most popular specification in peoples' mind, similar to the horsepower of a super sports car. The pressure alone which enables (and which is the price to be paid for) the use of highly efficient UHPLC stationary phase materials based on sub-2-μm particles does not give you automatically a fast and highly efficient separation. The whole design of the UHPLC system fluidics, that is, all volume segments in touch with the mobile phase or the sample, needs to be dressed to the requirements of these maximum-efficiency UHPLC materials. Two fluidic characteristics of such systems are in the primary focus here, the *gradient delay volume* and the *extra-column volume* (Figure 2.1).

The *gradient delay volume* or *GDV* is defined as the sum of all volume contributions from the point of gradient formation by blending of different solvents to the column head. Hence, the GDV has a major impact on the appearance of a chromatographic gradient separation; it is the reason for any gradient separation to begin with an isocratic hold-up step which takes as long as a change in the mobile phase composition needs to reach the column head and to interfere with the separation process. This implies also a critical limitation for method speed-up, depending on the selected flow rate of the separation.

The *extra-column volume* or *ECV*, however, is formed by the sum of all system volume contributions the sample needs to travel through, excluding the void volume of the separation column itself. This means that all volume segments from the position of the sample plug in the injector loop to the detector recording the separation result add up to the ECV. The ECV has much less of an impact on the speed of analysis, but it substantially determines the quality of the resulting chromatogram. All volumes the sample needs to pass make the sample zone and, thus, the peak in the chromatogram broader, caused by diffusion and other dispersive impacts like flush-out effects. The more efficiently the separation column works, which means the sharper and slimmer the resulting peaks are, the more important it is to minimize the ECV; otherwise, the quality of the separation *in* the column will be partially ruined by dispersion *outside* the column.

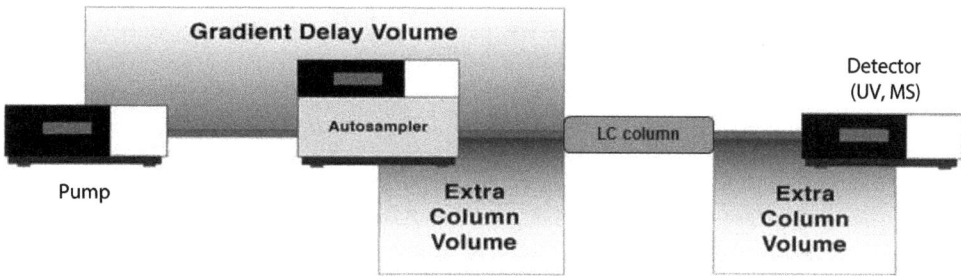

Figure 2.1 Illustration of the gradient delay volume (GDV) and the extra-column volume (ECV) of an LC system.

2.1.2.1 Speed in LC/MS Analysis I: Struggling with the Gradient Delay

High-throughput screening is one focus area for many LC/MS applications, for instance, in drug research and development in the pharmaceutical industry. Analysis times of less than 2–5 min for samples of modest complexity enable the fast and reliable processing even of large sample pipelines in an uninterrupted 24/7 routine operation, which makes this approach highly attractive, for example, for combinatorial synthesis monitoring or drug metabolism and pharmacokinetics studies (DMPK). With such short analysis cycles, the gradient delay volume (GDV) of a UHPLC system becomes a critical factor for the overall sample throughput. LC/MS applications in particular ask for separation columns with small inner diameters (from 2.1 mm I.D. columns for analytical scale separations down to several dozens or hundreds of microns in nano- and cap-LC applications), which come along with downscaled flow rates of less than 1 mL/min, with typical values between 50 and 500 µL/min. A small GDV is of great advantage here: The fastest gradient program is useless if a GDV of 500 µL in combination with a flow rate of 500 µL/min makes the changed eluent composition arrive at the UHPLC column head with a delay of one full minute. And please do not get blinded by smart marketing messages of many instrument vendors, which in most cases only specify the mixer size of the (U)HPLC pump: of course the mixing volume is part of the GDV, but the total GDV amount is much more than that; it includes the sample loop and other fluidic parts of the autosampler as well as all connecting tubing or, for instance, the whole pump head fluidics in case you are using a low-pressure gradient pump (LPG).

Pump Type and Mixing Volume

Pretty much all modern (U)HPLC pumps allow you to realize an overall GDV of 250 µL or less – getting much below 100 µL of GDV, however, is still a major challenge. Due to their operation principle, *high-pressure gradient pumps (HPG)* have an inherent advantage with respect to GDV compared to *low-pressure gradient pumps (LPG)*; this makes a HPG pump the preferred solution particularly for LC/MS applications. Using a LPG limits your LC method speed-up capabilities for LC and LC/MS separations significantly; this can only be overcome by reducing all potential GDV contributions, for example, by installing a smaller eluent mixer. But: wouldn't the mixing efficiency suffer from such a mixer change? Well, it depends on which perturbation effect would be affected, the *radial mixing* (i.e., along the cross section of your fluidics) or the *longitudinal or axial mixing* (along the flow direction) of your mobile phase. Radial mixing is usually achieved by complex shifts and changes in the liquid stream, for instance induced by a mixing helix or by branched channel structures on a chip-design mixer. Radial mixing is most required by HPG pumps due to their operation principle; fortunate enough this needs only a small mixing volume. Axial mixing in contrast is most effectively achieved by adding larger volume segments, like large-bore tubing or cylinders. Unfortunately, it is the LPG pump operation principle that asks mostly for axial mixing. As a consequence, reducing the volume of your mixing device will have much less of an impact on the performance of a HPG than of a LPG. In addition, baseline stability, drift, and noise suffer less from an axial inhomogeneity of the mobile phase in MS detection than in UV detection. All these are good arguments that a small mixing volume combined with a HPG pump is one key to success for high-throughput LC/MS analyses.

The GDV discussion is a particularly difficult one for pumps still having membrane-based pulsation dampers. In this case, the GDV also depends on the system pressure, and by that on the separation flow rate [1], mobile phase composition, temperature, and separation column type. While all major manufacturers of modern UHPLC pumps nowadays use electronic control mechanisms in their high-end and most middle-class instruments that allow a virtually ripple-free flow delivery without dampeners, some older pumps or simpler entry-level models still have to rely on mechanical pulse dampening. Using such a pump technology together with MS detection is, therefore, not advisable in general.

But what to do if you cannot further reduce the GDV of your system but still want to profit from a very short and steep, a *ballistic* gradient separation? Well, a workaround can be a delayed sample injection. It sounds simple: your autosampler injects the sample not simultaneously with the pump starting the gradient program, but the sample is introduced with a certain time delay which equals the GDV to be saved at the programmed flow rate. This operation principle is especially applicable if you need to transfer a separation method coming from a system with a lower GDV than your target system, as it allows you to reduce the effective isocratic hold-up the sample sees after injection. However, this is beneficial if you look on one single chromatogram, as it reduces the overall data acquisition time for this run: data recording still starts with the time of injection, not with the start of the pump program. But the total run time for this separation, your cycle time, still remains the same; however, you can reduce the column equilibration time at the end of the separation by the delay time of the injection because you equilibrate longer with the higher GDV before arrival of the gradient. So delayed injection is a nice workaround for method transfer, but it will not help you to substantially increase your sample throughput.

Sample Injection
Another important contributor to the GDV is the autosampler, which offers a lot of optimization potential. Users can typically choose between different sample loop sizes (read: GDV contributions); the default sample loop and system tubing are typically selected in a way that they universally cover pretty much every injection volume from fractions of a microliter to up to 100 µL volumes and more. This requires sample loops of significantly more than 100 µL internal volume. Most UHPLC-MS separations, however, run on 2.1 mm I.D. columns (or less, see also Section 2.2.1.1) and work with much less than 10 µL injection volume to avoid volume and/or mass overloading of the stationary phase. Cutting your sample loop size from nominal 100 µL to less than 30 µL will also reduce the GDV contribution of the autosampler accordingly. Autosamplers with the injection needle being part of the sample loop (*split-loop* principle, also *Flow-through-needle* principle) may additionally benefit from a small-sized needle seat capillary. If your system comes

with a motorized high-pressure syringe as part of the sample loop (a *metering device*, as realized, e.g., by Agilent Technologies and Thermo Scientific), then this will also contribute to the system GDV. Smart settings for the metering device help here to reduce this contribution to the system GDV. The Vanquish UHPLC systems from Thermo Scientific offer users to modify the GDV setting by a variable metering device piston positioning which allows a flexible adaptation of the autosampler GDV contribution to your LC separation – a feature which is particularly beneficial for method transfer. And last but not least, many instrument control software programs offer a bypass mode for autosamplers which optionally turns the injection valve back to the "load" position after injection. This eliminates the sample loop contribution to the GDV for the rest of the run – a quite significant amount for all split-loop autosamplers. One drawback on this feature is that it cuts a certain volume segment out of a running gradient program, which may have adversary effects on the separation and which needs to be considered during method development; we will discuss the pros and cons of this feature again in Section 2.1.2.3.

System Tubing

A factor which is frequently more over- than underrated (in contrast to the topics previously discussed) is the GDV contribution of the system tubing between the pump and the LC column. We will discuss in Section 2.1.2.2 that for band-broadening reasons, the shortest possible connection tubing between LC outlet and MS inlet is recommended, which leads to the some preference for a bottom-up LC stack design. However, the ascending flow path of such a bottom-up installation of a modular (U)HPLC system leads to potentially longer connection capillaries before the LC column than in a conventional top-down setup (if, for instance, the degasser is not integrated in the pump module, as exemplified in Figure 2.2b). But no worries – even if a connection line of 0.18 mm I.D. has a length of 19.7 inch/500 mm, this tube will have "only" 15 µL in volume, which is typically much less than 10% of the total system GDV. You might think that these 15 µL, however, could potentially harm much more as a contribution to band broadening by extra-column volumes. Well, this depends on where this tubing is placed. Whenever a larger bore capillary is installed in front of the autosampler, the sample does not encounter it anyway, and peak broadening is no issue at all. And even with wider capillaries sitting between sample loop and column head – the overwhelming majority of LC/MS separations are run in gradient mode. Due to the sample re-focusing effect of the gradient program which enriches the analytes on the column head by a huge initial retention at the very low starting solvent strength (*peak compression*), the impact of band broadening volumes in front of the LC column is massively reduced. Hence, any capillary tubing affects noticeably neither the gradient delay nor the band broadening. This statement, however, is only valid for analytical scale LC separations – things look different for capillary and nano-LC applications.

Take-home messages

- Minimize the gradient delay volume (GDV) of your LC system – it will help to remarkably reduce analysis time. High-pressure gradient pumps are of inherent advantage here
- GDV means more than the pump mixer. Assess all volumes from the gradient formation point to the column head for minimum GDV, but without sacrificing mixing performance
- Contributions of connection tubing to the GDV can typically be neglected in case of analytical scale LC separations.

2.1.2.2 Extra-column Volumes

As with UHPLC standalone installations, also LC/MS hyphenated systems are significantly prone to extra-column volume (ECV) contributions. A general rule of thumb says that the maximum ECV between sample introduction and point of detection should not exceed 10–15% of the peak volume of an eluting sample zone. A quick calculation illustrates the situation: a compound eluting in a 10-s wide peak (baseline width) – a level easily undercut by superfast UHPLC separations which can provide less than sub-2-s baseline widths – has a peak volume of 83 µL at a flow rate of 500 µL/min. This translates in a tolerable extra-column volume of only 8–10 µL. This is even more a challenge in LC/MS, as here also the bridge between the UHPLC outlet and the mass spectrometer inlet contributes to the ECV. Thus, this bridging tubing ideally would have the smallest volume you can think of, just to minimize unwanted band broadening effects. We can achieve this quite easily by using a capillary of a very small I.D. which should also be as short as possible. However, this capillary cannot be infinitely short – due to the physical dimensions and the geometric arrangement of UHPLC and MS instruments there will always be a certain minimum distance that you will need to bridge. Simultaneously, slim capillaries always generate high system backpressures – remember Hagen–Poiseuille's law (Eq. 2.1) which describes the capillary pressure as being inversely proportional to the 4th power of the capillary radius:

$$F = \frac{V}{t} = \frac{\Delta p \cdot \pi \cdot r^4}{8 \cdot \eta \cdot L} \qquad (2.1)$$

where F = flow rate, V = volume, t = time, Δp = pressure difference, r = capillary radius, η = fluid viscosity, L = capillary length.

Considering this, we can deduce three general recommendations.

(1) Install Your (U)HPLC System in a Smart Way

Reducing the pathway length between LC and MS starts already with setting up your UHPLC instrumentation. The conventional LC setup typically follows a top-down path of your mobile phase (Figure 2.2a): with the solvent bottles on top, the stack sequentially contains the degasser, the pump, the autosampler, the column thermostat, and finally the detector(s) downstream. Most of all commercial mass spectrometers, however, have the ion source inlet placed in a given height

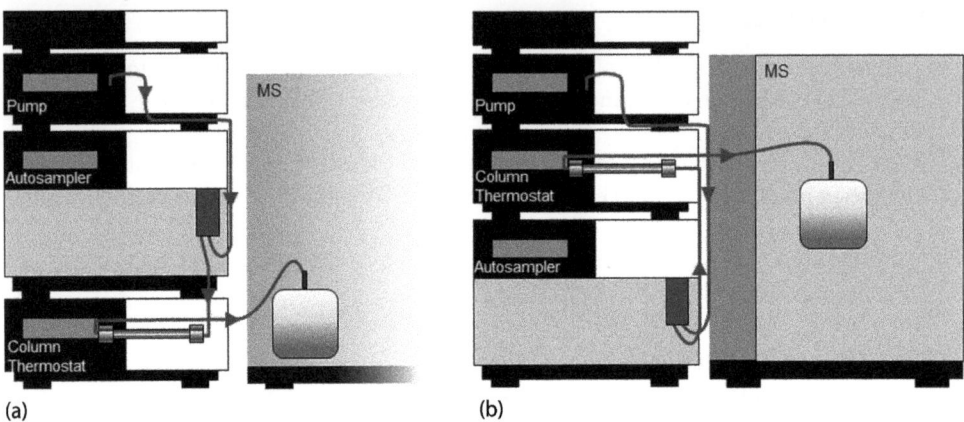

Figure 2.2 Descending top-down (a) and ascending bottom-up (b) flow path for minimized connection tubing length between LC column outlet and MS inlet (ion source).

above benchtop level, typically between 12 and 25 inches (300–600 mm). Hence, a bottom-up installation of the UHPLC flow path would be more appropriate – ideally with the pump and degasser at the bottom, then upward followed by the autosampler, the column thermostat, and – if required – in-line detectors (Figure 2.2b). Most modern compact instruments are already designed like that by default. Modular LC systems can be individually configured by the user, allowing for a reduction of the LC-MS connection capillary length by up to 8–12 inches (200–300 mm) in a bottom-up setup compared to a conventional top-down installation. This, for instance, translates into a 2.4 µL void volume saving when using a 100 µm I.D. connection tubing. In some cases, the bottom-up setup may come along with a slightly longer tubing in front of the LC column; however, as already discussed in Section 2.1.2.1, this does not noticeably impact the separation quality in case of a gradient separation.

(2) Keep Your Connection Tubing Slim and Short

Reducing the internal tubing volume always goes in line with short lengths and small I.D.s. Thus, the connection tubing between your LC outlet (either the column or, if present, an additional detector, e.g., a diode array detector) and your MS inlet (the ion source) should have the smallest inner diameter possible which does not use up too much of your (U)HPLC system pressure capabilities and which does not compromise the pressure stability of any part of the flow path prior to the connection line (e.g., UV detection flow cells). As an example, take a connection line of 0.13 mm I.D., having a length of 30 inches (750 mm): running this capillary at 25 °C and a flow rate of 500 µL/min would generate a backpressure of moderate 11 bar (160 psi) for a mobile phase with a viscosity of $1.2 \cdot 10^{-4}$ Pa s (which is slightly more than the viscosity maximum of water/acetonitrile mixtures at ambient temperature). However, this capillary would contribute 10 µL of extra-column volume *behind* your LC column, where it is particularly critical. Convert-

Table 2.1 Volumes and backpressure of a 30 inches/750 mm capillary with different inner diameters (I.D.) in the viscosity maximum of water/acetonitrile and water/methanol mixtures.

		0.13 mm I.D.	0.10 mm I.D.	0.075 mm I.D.
	Volume	10.0 µL	5.9 µL	3.3 µL
Water/acetonitrile 91/9 v/v,	Pressure at	9.5 bar/	27 bar/	85 bar/
$\eta = 1.06$ cP at 25 °C	$F = 0.5$ mL/min	138 psi	392 psi	1233 psi
	Pressure at	19 bar/	54 bar/	171 bar/
	$F = 1.0$ mL/min	276 psi	783 psi	2480 psi
Water/methanol 40/60 v/v,	Pressure at	14 bar/	40 bar/	125 bar/
$\eta = 1.56$ cP at 25 °C	$F = 0.5$ mL/min	203 psi	580 psi	1813 psi
	Pressure at	28 bar/	80 bar/	251 bar/
	$F = 1.0$ mL/min	406 psi	1160 psi	3640 psi

ing this tubing to 0.10 mm I.D. reduces the extra-column volume contribution to 5.9 µL, but it comes along with a rise in pressure to 31 bar (450 psi). A 0.075 mm I.D. capillary reduces the void volume contribution further down to 3.3 µL, but at the cost of a considerably high backpressure of 97 bar (1410 psi). For further illustration, Table 2.1 summarizes some model calculations for typical LC/MS application conditions. As we can see, a significant speed-up of LC separations at flow rates beyond 1 mL/min is barely possible (leaving out the question if this was useful with respect to the MS detection sensitivity at such high flow rates).

In case you have an additional detector in front of your mass spectrometer, like, for instance, a UV detector, you also have to take care of the detector flow cell pressure limit. Depending on the design principle, the maximum pressure limit of commercial UV flow cells can vary between 870 and 4350 psi (60 and 300 bar). Please be aware that it is not only the MS connection tubing that generates an additional pressure load to your UV flow cell; many mass spectrometers use internal switching valves to introduce calibrant solutions into the MS ion source, which block the flow path completely for a fraction of a second when they are actuated, thus, generating a very short but also very high pressure spike to any technical part in the flow path in front of it. In case you have a more fragile detector cell, you may want to consider either to split your LC column effluent by a tee piece, or to bypass the MS switching valve by directly connecting your LC system with the ion source sprayer. The former reduces peak efficiencies (by the band-spreading tee piece connection) and sensitivity (only the split fraction of your effluent runs into the MS detector), the latter even tends to improve your peak efficiency in the MS chromatogram, as switching valves in general are plate count killers due to their large bore and groove sizes; however, bridging the internal MS switching valve forbids you to re-calibrate your mass spectrometer automatically in a sequence run. How critical this is depends on the mass spectrometer type; as we will see in Section 2.1.3, some mass spectrometers need a frequent, if not permanent calibrant infusion, while others do not.

Just as a concluding note – although the tubing I.D. now may be seen as problematic due to its huge impact on the system pressure, it is by far the smarter optimization parameter to reduce volume contributions. The I.D. goes with the volume by the second power, while the length contributes only linearly to it. For comparison reasons, let's take again our 30 inches (750 mm) long capillary of 0.13 mm ID (a quite common example for connecting LC with MS) which has an internal volume of approximately 10 µL. To reduce this by half, you would need to cut the tube down to half the length, so 12.8 inches (375 mm) – which might be too short to make your LC/MS connection. Changing to an I.D. of only 0.10 mm (−23%) brings you down to a volume of 5.9 µL, which comes close to the reduction by factor 2, but preserves your original capillary length so that you still have a good chance to be in line with your instrument arrangement. Alternatively, going down to a 0.10 mm I.D. capillary would enable you to make the capillary $(0.13/0.10)^2 = 1.7\times$ longer (i.e., 43.4 inches/1275 mm) but still keeping the same internal volume of 10 µL. Looking at the impact the tubing volume has on the peak volume, the picture becomes even more critical, as volume is not equal to volume. The peak volume only increases with the *square root* of the tubing *length*, but with the *square* of the tubing *diameter*. Comparing adding length with increasing the I.D., an I.D. change is therefore even to the power of 4 as severe as a length variation. We see clearly now: for extra-column and gradient delay volume, capillary I.D. rules over length.

(3) Take Care of Your Fitting and Tubing Connection Quality

It is not only the hold-up volume of your tubing that matters. The quality of your fitting system also has a large impact on the quality of your LC/MS chromatogram – a factor which is typically underrated in everyday lab life. It is still common use today to connect a UHPLC system with a mass spectrometer by cutting a PEEK tube of an appropriate length from the bulk and installing it using PEEK fingertight fittings. However, due to improper cutting quality, nonrectangular tubing ends and careless fitting fixing, this introduces a measurable but pointless and avoidable void volume that can be a real efficiency killer. Various dead-volume reducing fitting systems are nowadays commercialized; however, most of them are specially designed and tailored to a respective (U)HPLC equipment and, thus, not universally applicable. Only four universal UHPLC fitting systems are currently available on the market, not all of them also covering true UHPLC pressure loads. These are the Viper™ fingertight fitting technology from Thermo Scientific [2] (max. pressure of up to 22 000 psi/1500 bar), the A-Line™ fitting design from Agilent Technologies [3] (max. pressure of up to 18 850 psi/1300 bar), Sure-Fit™ from MicroSolv Technology Corporation [4] (now IDEX; max. pressure of up to 6000 psi/413 bar), and MarvelX™, also from IDEX (up to 19 000 psi/1310 bar) [5]. Figure 2.3 illustrates how using such virtually zero-dead volume connections provide a remarkable gain in plate numbers and resolution. The chromatogram in Figure 2.3a was generated using a standard PEEK capillary of 0.13 mm I.D., cut from the bulk by a standard tube cutter and installed between LC column outlet and MS ion source inlet by regular PEEK

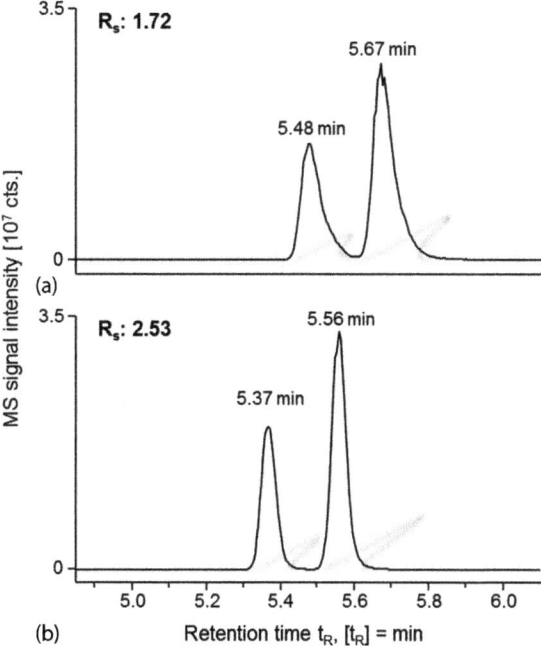

Figure 2.3 LC/MS chromatogram of two isomers, $m/z = 240.10$; (a) PEEK bulk capillary behind the column (0.13 mm I.D.), PEEK fingertight fittings; (b) SST capillary with virtually zero-dead volume connection behind the column (Viper™ fingertight fitting technology, 0.13 mm I.D.)

fingertight fittings. The chromatogram in Figure 2.3b was run under exactly the same conditions, the only difference being a Viper stainless steel (SST) capillary of identical size and dimension of the PEEK one between column and ion source. The significant rise in chromatographic resolution of 47% from 1.72 to 2.53 clearly illustrates how much separation power is wasted in most LC/MS installations simply due to the use of improper tubing and fitting quality.

Take-home messages

- Keep your eyes on the shortest distance possible between LC outlet and MS inlet already while setting up your UPHPLC system
- Do not worry too much about the extra-column volume in front of the LC separation column – it can typically be ignored in gradient elution mode thanks to a peak compression effect
- Focus on the ECV behind the LC column instead: shorter and slimmer tubing always pays off. However, take care of the backpressure generated by very thin capillaries – they eat up pressure reserves of your UHPLC system and could potentially kill your UV flow cell
- Get rid of any uncontrolled extra-column volume contributions due to improper tubing cuts and connections by using state-of-the-art zero-dead-volume fitting systems.

2.1.2.3 Speed in LC/MS Analysis II: The Total Cycle Time or How Fast Can I Be?

As already discussed earlier for the *delayed injection*, it is not the speed of your LC separation alone which matters for the total cycle time. Various other actions add up to it here, including every step of liquid handling like sample aspiration, needle wash cycles, or column re-equilibration at the end of your separation.

The first bottleneck for speeding up the total cycle time is already the preparation of the sample injection, as this is fairly time-consuming and also depends on various instrumental properties. Fast state-of-the-art autosamplers can realize injection cycle times of less than 10–30 s. This impressive speed, however, can only be achieved with very high sample draw speed and without any external needle wash steps. So there is a price to be paid: too fast a sample aspiration negatively affects the injection precision, especially with viscous samples or low-boiling sample solvents, while not cleaning the exterior of the injector needle will lead to enhanced carryover effects. Some UHPLC control software programs offer a "prepare next injection" feature which already initiates drawing a new sample into the bypassed sample loop while the previous LC separation run is about to be finished. This partial parallelization of injection preparation and chromatographic separation indeed leads to a shortened total cycle time and still gives sufficient time for a thorough wash procedure of the needle unit. However, such an interlacing of injection and analysis steps always requires the sample loop being switched off the fluidic path by turning the injection valve from the *Inject* back to the *Load* position at some point in your LC separation. With modern split-loop samplers, the sample loop is a permanent part of the fluidic path, thus, ensuring low sample carryover due to a continuous loop rinse by the mobile phase. Here it is essential that the switch-back takes place only when the sample loop is filled with your initial mobile phase composition. Otherwise, this will cut a certain volume segment out of your gradient profile, thus, actively interfering with the chromatographic process running in your LC column. Hence, an ideal point in time to trigger all kinds of bypassing actions would be during the column re-equilibration step at the gradient end; the only prerequisite here is that the bypass actions do not negatively affect the chemical equilibration of the column. However, if for whatever reason you cannot make use of such a parallelization of sample aspiration and running separation, it is barely possible in real life to achieve a precise and an ultralow carryover injection in less than 30-s total cycle time.

Let us next look at the injection process to identify further optimization points. The time needed to fill the sample loop can be optimized both by the piston speed of the injection device (typically a glass syringe or a high-pressure piston) and by the injection volume. Thanks to the small injection volumes in UHPLC of less than 1–5 µL, a moderate piston draw speed of 250 nL/s still ensures a rapid though reproducible sample dosage even for viscous samples or highly volatile sample solvents. But liquid handling is more than only drawing and injecting dissolved samples. Cleaning internal parts of the sampler fluidics that are in touch with the sample but not continuously flushed by the mobile phase can also become a time-critical step. Some instrument hardware designs use an injection syringe for these internal rinsing steps, and so this cleaning takes longer the smaller this syringe volume is. Washing a sampler tubing of, for instance, 40 µL with only the fourfold volume of 160 µL wash liquid can take a considerable amount of time, and a 100 µL syringe finishes this cleaning obviously 4 times faster than a 25 µL syringe. With very unfavorable settings, for instance when large tubing volume, small cleaning piston volume, and very fast LC separations of less than 2 min of run time come together, cleaning the autosampler fluidics can even take longer than the entire analytical separation.

But it is not only before or at the beginning of your separation where you have the potential for cycle time optimization. There is also one time-burner at the end of your chromatography, and it can be a substantial one: it is the column re-equilibration. When running a gradient separation, it is imperative to re-condition the stationary phase back to the initial mobile phase composition of the gradient once the solvent strength gradient has reached its final level. This is the only way to ensure the mobile and stationary phase being in an equilibrium state, which is a prerequisite for stable retention times. As a rule of thumb, it is recommended to flush the separation column with at least the fivefold column void volume V_m of mobile phase for a stable equilibrium state. With challenging analysis conditions, the required equilibration volume can easily go up to 8–10 times of the void volume; This is frequently the case either at low initial organic solvent amounts of less than 5% or with analytes strongly affected already by minor deviations from the equilibration state – typically observed for analytes with retention factors of $k < 1$ or for pH-sensitive separations. A short example shall illustrate the time impact here. We will calculate the column void volume V_M from the geometrical column volume V_C using Eq. 2.2:

$$V_M = \varepsilon_t \cdot V_C \tag{2.2}$$

where ε_t = total porosity, r = column radius, L = column length, and $V_C = \pi r^2 \cdot L$.

Table 2.2 lists two different use cases for comparison, a fairly short UHPLC column for high-throughput screening (HTS), and a long column for a high-resolution analysis, both columns operated at 500 µL/min, which is a good average for a sub-3-µm packing material and is still MS compatible. You will immediately see that even the HTS column of 2.1 × 50 mm and a typical total porosity of $\varepsilon_t = 0.65$ needs a re-equilibration time of 1.1–1.8 min. The five times longer high-resolution column consequently will require the fivefold of re-conditioning

Table 2.2 Recommended re-equilibration volume for high-throughput and high-resolution columns under typical MS-compatible conditions.

Column dimension ID × L (mm)	V_M	Required re-equilibration volume (rounded)	Flushing time for 5–8 · V_M at 0.5 mL/min
2.1 × 50	113 µL	570–900 µL	1.1–1.8 min
2.1 × 250	563 µL	2800–4500 µL	5.6–9.0 min

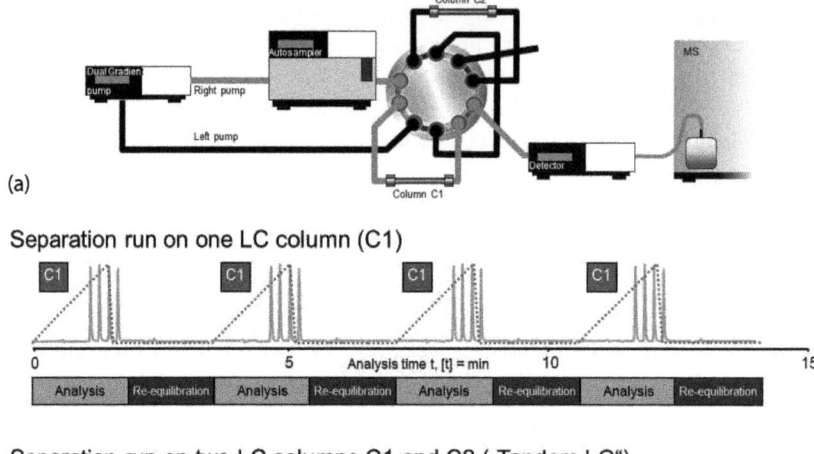

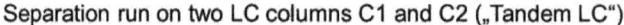

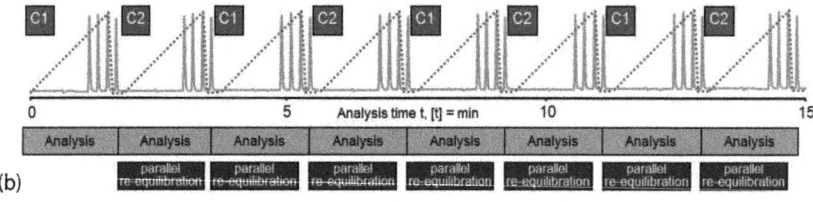

Figure 2.4 Reduction of re-equilibration time and throughput enhancement by using a second separation column and alternating sample injection (Tandem LC); (a) flow scheme, (b) injection interlacing.

time, ending up at something between 5.5 and 9 min. With any regular UHPLC system, this amount of time adds to each and every single injection, no matter how fast the gradient separation itself will be. Hence, a total cycle time of less than 2 min is barely achievable. The only way to solve this dilemma would be a second separation column of identical properties to the original one, which could be equilibrated in parallel to a running analysis using a second pump and a suitable switching valve. The literature usually refers to this setup as Tandem LC. Once the analyte elution on one column has finished, the next injection is then done alternatingly on the other column, with the previous column being washed and equilibrated simultaneously (Figure 2.4).

Finally concluding these considerations on side procedures of (U)HPLC separations will leave us with one quite sobering finding. A fast separation method alone is by far no guarantee for a high sample throughput and short cycle times. Typically, the most time-burning process is the column equilibration which in most cases can only be shortened at the expense of reproducibility. If we do not consider injection interlacing steps, then any high-throughput UHPLC analysis is on average extended by 0.5 min for preparing the sample injection and by 1.5 min for column reconditioning, with longer times required easily, depending on sample type, potential wash cycles, and column dimensions. A considerably fast LC method with a 2-min separation time then quickly doubles the cycle time, and even an injection preparation parallel to the final phase of a running analysis does not help substantially to shorten the total cycle time. A minimum total cycle time of much less than 4–5 min is therefore hard to beat, even with the fastest separation programs on the most advanced UHPLC equipment.

Take-home messages

- A rapid separation means much more than an ultra-short (ballistic) gradient program
- Fast autosamplers help to shorten the delay time prior to the sample injection. Preparing the next injection while the current analysis is still ongoing can help to additionally reduce the cycle time
- Column re-equilibration is a time-burner, which, however, is hard to avoid, as a thorough column equilibration is mandatory for robust separation results.

2.1.3
Does Your Mass Spectrometer Fit Your Purpose?

It is a long-stressed platitude that the right tool makes all the difference: anyone who has ever tried to fix an inch hex bolt with a metric wrench will confirm this from personal experience. Well, what applies to screwdrivers is in fact not different to high-tech analysis equipment in your lab, and it is particularly correct for mass spectrometry. Currently, five different mass analyzer principles are established on the market for LC/MS applications:

- Quadrupole (Q)
- Ion trap (QIT, LIT)
- Time of flight (TOF)
- Orbitrap
- Ion cyclotron resonance (ICR).

Pretty much all commercial LC/MS instrumentations rely on (at least) one of those five mass selectors; more sophisticated devices may either vary slightly in their technical design (e.g., 3D or Paul trap, *QIT*, versus linear ion trap, *LIT*), or come as hybrid instruments combining two or more of these analyzer types (e.g., triple quadrupole, *QqQ*, quadrupole fime-of-flight, *Qq-TOF*, ion trap Orbitrap, *LIT*-Orbitrap, or even Tribrids merging three different analyzers into one device). Each of those solutions has its strengths and weaknesses which make it more appropriate for certain applications than for others. The Chapters 1 and 3 of this textbook give a comprehensive overview on the technological state of the art; for additional information refer also to [6, 7].

But whatever field of application you are looking into – pretty much every analytical challenge requiring mass spectrometric detection can be reduced to either one of the two aspects:

- Selective detection of previously known analytes with highest sensitivity for quantitation, *or*
- Identification and structure elucidation of unknown compounds.

Combining these tasks with the technical potential of UHPLC which enables ultra-high separation performance and/or high speed of analysis will then result in a very attractive technology for the fast *and* comprehensive screening of complex samples with low sample preparation efforts (*dilute-and-shoot*) and high throughput. However, the capabilities of a mass spectrometer need to keep pace with the rising requirements dictated by higher sample complexity and shorter analysis times. You will get the highest confidence in your result if you apply the most suited mass spectrometer to a given analytical problem. Let us briefly discuss the pros and the cons of the various mass spectrometer types for our two core analytical tasks from above, either the quantitation of known analytes as specific and sensitive as possible (*targeted screening*), or the fishing in the troubled water of samples where you do not have a clue about what compounds to expect (*screening for unknowns*).

For **targeted screening**, with a clear focus on quantitation of previously known target analytes, all those MS types are preferred that combine two mass analyzers with a collision cell in-between, allowing for *collision-induced dissociation (CID)* by *tandem-MS in space*. From all potential MS/MS operation modes offered by these instrument types, targeted screening is most frequently run in the *Selected Reaction Monitoring* mode (*SRM*, also called *Multiple Reaction Monitoring, MRM*). This operation mode requires that you have a good understanding of how your target analyte dissociates into characteristic and ideally specific fragments after exciting it to vibrations by collision with an inert gas in a collision cell. For the collision gas, the heavier argon is typically preferred over the lighter nitrogen for a higher kinetic impact. You will operate the two mass analyzers as ion filters then; the one in front of the reagent cell eliminates all unwanted ions so that only the ions with a *m/z* value of your target analyte, the *precursor ions*, enter the collision cell. The mass filter behind the cell then is set to the *m/z* value(s) of the expected *fragment ions*. This SRM operation mode features two main advantages: the combination of a precursor ion with as many characteristic fragment ions as possible substantially increases the detection specificity, and it ensures tremendously low limits of detection (LoD) and quantitation. Running the MS in SRM mode not only filters out all unwanted interfering ions, thus, virtually eliminating baseline noise; it also reserves the full MS duty cycle exclusively for the detection of the target analyte ions, allowing you to detect a much higher amount of your target ions than in a full-scan mode. Up to now, triple quadrupole mass spectrometers (QqQ) are the uncrowned leaders in the targeted screening domain, being superior to Qq-TOF or other instrument types with respect to sensitivity, ease of use, result robustness, and profitability. Particularly ion trap mass spectrometers which basically offer the inherent advantage of *tandem MS and MS^n in time* for even more specific fragmentation experiments are not ideal for quantitation purposes due to their limited linear detection range (refer also to the *space charge* phenomenon in Section 2.3.5). Additionally, due to their operation principle ion traps will completely fail for all MS/MS operation modes that require a scan process as the first step in a series of MS experiments. If you need to perform a "true" *precursor ion scan* or a *constant neutral loss scan* for your analysis (and not merely a software dataset reconstructing these scan modes out of a set of sequential MS^n experiments), then the use of a *tandem MS in space* machine such as a triple quad device is imperative. It should be noted that depending on the molecular mass of your target analytes, the preferred MS/MS instrument type may slightly vary. For small molecules of typically less than 1000–1200 Da, a triple quad machine clearly rules out other MS types due to its benefits in robustness, sensitivity, and investment costs. However, the comparably low upper m/z limit of QqQs may be of slight disadvantage; Qq-TOF and Orbitrap instruments are more in favor for large and macromolecules.

The other focus for LC/MS applications is the **screening for unknowns**, where you primarily need to learn about unknown sample constituents as much as you can with a very low experimental effort – ideally within one single LC/MS injection. The most relevant information you would need to gather comprises

- *The elemental composition* – which can be derived from high resolution/ accurate mass (HR/AM) mass measurements
- *Molecular substructures* – to be determined by MS/MS or MS^n experiments and
- *Signal intensity ratio of the isotopes*, the so-called isotope pattern, which backs up the elemental composition calculation based on HR/AM results.

As already discussed in Chapter 1, only time-of-flight (TOF), Orbitrap, and ion cyclotron resonance (FTICR) mass analyzers allow for reliable HR/AM measurements with a sufficient mass accuracy of less than 5 ppm and resolving power. Coupling these analyzers with a quadrupole and a collision cell upfront enables you to additionally measure CID fragment spectra revealing details on molecular substructures, functional groups, etc., thus, supporting structure elucidation. Data acquisition speed and mass resolving power R behave strictly opposite within these three MS types: as of today, TOF devices are by far the fastest mass spectrometers on the market (max. scan rate of up to 200 Hz), followed by Orbitraps (up to 20 Hz) and FTICR (1 Hz or less). In contrast, FTICRs lead in resolving power (R up to 10 000 000), followed by Orbitraps (R up to 500 000) and TOFs (R up to 80 000).

- *TOF devices* offer exciting scan speeds, high mass accuracy, and resolving power at a good price per performance; however, they tend to be very prone even to minor variations of the environmental conditions. As with all materials, also the flight tube of a TOF MS expands with higher temperature. An elongation (or shrinking) of the flight tube even only on the micrometer scale will substantially affect the accuracy of the mass determination (to be precise: the mass/charge determination) and the resolving power. For a stable and rugged experimental result, you will need a powerful and precise air conditioning in your MS lab (be also aware of sun glare shining on the mass spectrometer through the windows!) as well as a regular mass calibration, approximately once per hour or even more frequently, to compensate for any drifts. As a drifting mass axis calibration can easily occur already on the timescale of one LC separation, the highest confidence in your mass accuracy can only be guaranteed by an internal mass calibration where known mass calibration compounds are permanently co-infused into the MS during the LC run. Some TOF devices offer a continuous calibrant infusion as a *lock spray* into the ion source using a revolving aperture that alternatingly passes either the LC effluent or the calibrant solution into the mass analyzer. As an alternative, the calibrant solution can also be added to the LC effluent in front of the ion source by a simple tee piece setup. For a correct data analysis, every measured m/z value taken from the LC/MS dataset is then referenced against the m/z values of the known calibrants. This may sound a bit clumsy, but it is the only way for TOF devices to fully reach their maximum specified mass accuracy.

- *Orbitrap devices*, in contrast, are much more rugged against changes in the ambient conditions due to their inherently different design and operation principle. For routine applications, a mass calibration once a week is typically sufficient (depending on the application and lab conditions). Next to the higher analysis ruggedness, they are significantly superior to TOF devices in terms of mass resolution and at least par with respect to mass accuracy – in fact they are the only mass analyzers that come even close to the accuracy of FTICR but with much less challenging claims for technical infrastructure, as they are true benchtop instruments today.

- *FTICR instruments* are very, very sensitive, being capable to detect even down to 10 individual molecules within their detection cell; and they are unbeaten yet in terms of mass resolving power and mass accuracy. However, the very low data rate, the limited linear detection range, their bulky size, and last but not least the massive total costs of operation (think not only of the cost for the device alone but also of the demanding infrastructure for the superconductive magnet) will make this mass spectrometer type a highly dedicated expert system also in the foreseeable future, asking for a high level of user expertise and by that not having a real chance to establish in routine applications.

Ion traps (being the only MS type together with FTICR) featuring tandem MS in time and by that MS^n experiments with $n \geq 2$ are the most versatile instruments for substructure elucidation by gas phase fragmentation reactions. Due to their limited mass accuracy of typically higher than 10 ppm and only moderate resolving power, they are not really suitable for HR/AM analyses. Their preferred field of application is therefore the elucidation of analyte structures for compound classes with only a limited variability of building blocks, like the analysis of peptides, proteins, and nucleic acids.

A rather special position in the MS world is held by the fairly simple single quadrupole mass spectrometers. With their low mass accuracy (above 100 ppm) and quite poor resolving power (R about 1000 for $m/z = 1000$) they are neither good for structure elucidation/screening for unknowns nor for a targeted screening. Their strengths are robustness and a low price, and their mass results can at least support peak assignment during method development and serve as a negative confirmation on the absence of a compound of interest within the limit of detection. Therefore they are frequently used as screening detector for samples of low complexity, for instance in the open-access process control analysis of combinatorial reactions. Due to their limited mass spectrometric performance, many users do not even perceive single quads as true mass spectrometers but much more as mass-selective detectors (MSD), a concept which is meanwhile widely adopted by the marketing activities of various single quad manufacturers.

Table 2.3 gives a rough overview on the suitability of most common mass spectrometer types of today in combination with UHPLC for various application scenarios. In addition to the earlier discussed targeted screening and screening of unknowns, also more generalized aspects of structure elucidation and quantitative amount determination are listed here. It should be mentioned that this table has of course to live with a certain generalization. All major instrument manufacturers may offer individual, highly specialized flavors of the one or the other type of mass analyzer which exceeds the general limitations predicted by this list, but from a general perspective, this categorization applies very well to the different mass spectrometer capabilities and applicability.

Table 2.3 Suitability and purpose of various mass spectrometer types; + = well-suited, o = moderately suitable, – = inappropriate.

	Structure elucidation			Simple quantitation	Targeted screening
	Elemental composition	Determination of substructures	Screening for unknowns		
Q	–	–	–	+	o
QqQ	–	o	o	+	+
QIT	–	+	o	–	o
LIT	–	+	o	o	o
QTRAP	–	+	o	+	+
TOF	+	–	–	o	o
Qq-TOF	+	+	+	o	+
Orbitrap	+	o	o	+	o
Q-Orbitrap	+	+	+	+	+
LIT-Orbitrap	+	+	+	o	+
FTICR	+	+	–	o	–

2.1.4
Data Rates and Cycle Times of Modern Mass Spectrometers

It is common sense that an accurate quantitative result can only be generated from the best possible calculation of the *peak area* for your analyte of interest (and that of reference compounds of course); quantitation based on peak areas beats the peak height determination approach by far with respect to error deviation. Hereby, the higher the number of data points which scan the elution profile of your analyte, the smaller the error in peak area calculation and in deviation between the experimentally determined and the ideal, theoretical peak profile will be. To ensure an acceptably well recorded dataset, the measured chromatographic peak should at least be described by 25–30 data points. For classical LC detectors, this is no real challenge, as spectroscopic detectors (UV absorption, fluorescence) today provide data acquisition rates up to 250 Hz, which is more than enough to cope even with ultrafast UHPLC separations and peak widths in the 1-s range. Mass spectrometers, however, are by far not able to keep pace with the speed of state-of-the-art UV detectors. Moreover, data acquisition rates and duty cycles of mass spectrometers behave opposite to the data quality: in most cases, high data rates and short duty cycles come along with poor mass accuracy and reduced resolving power. Especially if the instrument needs to perform complex MS/MS or MS^n experiments, the duty cycles for the individual fragmentation experiments will take so much time that a high chromatographic data acquisition rate simply cannot be realized anymore: the mass spectrometer will be blind for a new package of the continuously infused ions as long as it processes the experimental steps of the previous set of ions. It is then up to you as the user to find the ideal balance between the requirements for high-quality *LC/MS chromatograms*, that is, a high data rate for best describing the concentration distribution of an eluting sample zone, and high-resolution *mass spectra* for high-confidence compound confirmation. The exact data acquisition rate of a mass spectrometer hereby depends on many different parameters: instrumental criteria like the mass analyzer type, the technical features and properties of your particular instrument such as electronics design, processor speed, but also on experimental conditions like the MS experiment type (is it run in full scan mode, in SIM or SRM mode, precursor ion scan, etc.), the actual mass range, or type and number of subsequent fragmentation steps (MS^n, data-dependent or data-independent MS/MS acquisition etc.). Today, time-of-flight (TOF) mass spectrometers represent the fastest mass analyzers with specified data acquisition rates of up to nominal 200 Hz for MS and 100 Hz for MS/MS runs [6]. This speed sounds very impressive, but it should be mentioned that this high speed does not allow simultaneously achieving also highest resolving power and spectrum quality. For comparison: triple quadrupole mass spectrometers which are the most widely used MS types for routine quantitation offer typical data acquisition rates of 5–15 Hz. This can already be challenged by a well-optimized conventional HPLC separation – to meet the requirements of ultrafast UHPLC separations with ballistic gradients, this will definitely be too slow.

2.1.5
Complementary Information by Additional Detectors or Mass Spectrometry Won't Save the World

It is a well-known saying that mass spectrometry is one of the most powerful analytical tools the world has ever seen. Without a shadow of a doubt, the sheer amount, the detail degree, and the accuracy of analytical information provided by mass spectrometers is very impressive; however, they cannot solve the impossible, and performing miracles beyond common sense is also not their business. Here is a small collection of the most widespread hypes and (partially) wrong assessments on mass spectrometers:

- "A mass spec is a universal detector."
  This is a frequently quoted claim, which, however, is not becoming more correct by frequent repetition. The advocates of this phrase typically compare MS with UV detection, referring to the fact that spectroscopic detection could only measure analytes having suitable chromophores which interact with electromagnetic waves of a certain energy (represented by the wavelength). This is perfectly right, but unfortunately it is only an indicator for the selectivity of UV absorbance detection, but not for the pretended universality of MS detection. Indeed, it ignores the fact that also the detectability in a mass spectrometer is analyte-dependent because it relates to the ionizability of your compound of interest; this, however, is not only related to the amount of ionizing energy present in the ion source, but also to analyte-specific properties. Molecules that do not have a considerable gas phase acidity or basicity will lead to a very poor ion yield in ESI or APCI mode, resulting in only a low amount of detectable molecular ions. Simply said: ESI and APCI are selective toward molecules with a certain gas phase acidity or basicity. Hence, every analyte species has its own individual mass spectrometric response factor, which indeed might be too low for a proper MS detection, depending on the selected ionization principle. Mass spectrometers are highly flexible in their wide application range and can easily be adapted to analyte requirements by a simple change of the ionization mode – but they are far away from being a universal detector.
  It should be mentioned here that in many conversations and also in some literature there is no clear distinction between a *universal* and a *uniform* detection. The latter one describes the requirement of providing a homogeneous, identical response factor for all analytes of interest, independent from their molecular properties. This does not necessarily has to come along with universal detection, but in real life it is an extended feature of virtually universal detectors. However, if a mass spectrometer does not detect universally by definition, there is even less of an argument for a mass spectrometer being a uniform detector, due to the different ionizability of the analytes. This implies that for all quantitation experiments, the mass spectrometer must be calibrated for each individual analyte – which comes along with a significant experimental effort. True universal detectors in contrast (or technologies that come close

to this ideal), for example, Charged Aerosol Detectors (CAD) or Evaporative Light Scattering Detectors (ELSD), can massively reduce (for exact quantitation) or even virtually eliminate the calibration efforts (for semiquantitative results and/or with constant matrix content).

- "There's no detector which is more sensitive than a mass spec."
  This phrase touches the same misapprehension as the previous one. Sensitivity and the limits of detection (LoD) and quantitation (LoQ) in mass spectrometry are not by default superior to any other detector. Under favorable conditions, like high ion formation yield and good ion transmission through the mass analyzer to the mass detector, mass spectrometers are indeed very powerful, allowing LoQs down to a femto- or even attomol level. However, in case of poorly ionizable analytes, an inappropriate ionization principle, and/or perhaps also not the most sensitive MS instrument design, there may be other detection principles that are clearly in favor, for instance, electrochemical or fluorescence detection.

- "Identify all your analytes with 100% certainty using a mass spec."
  Also this claim cannot be confirmed without limitations. Whatever fancy things you can do with your molecular ions in a modern mass spectrometer, like performing sophisticated gas phase experiments – at the end of the day, a mass spectrometer is "merely" a highly accurate type of balance to determine molecular masses. However, identical molecular masses by far do not equal identical molecular structures and appearances. Even the most accurate molecular mass determination will fail in distinguishing isobaric compounds entering the mass spectrometer at the same time, without further discrimination. Any kind of isobaric compounds – *E/Z* isomers, diastereomers, enantiomers etc. – cannot be differentiated by an MS experiment in the first place, as they all have the same molecular mass. As long as isobaric species show a different fragmentation behavior, you will have a certain chance to distinguish them based on their fragmentation pattern in an MS/MS spectrum; however, this is unfortunately not a given in many cases.

All these aspects clearly show that other, more classical detection principles are by far not obsolete only because you have a mass spectrometer in your lab. Especially when it comes to the structure elucidation of small molecules, spectroscopy (UV absorption, fluorescence) or electrochemistry provide valuable and complementary information helping to interpret your MS results – with NMR being the gold-standard of course, but LC-NMR hyphenation is by far more complex and less widespread than most other detection principles. One example to showcase the issue with isobaric compounds: many isomers differ significantly in their UV absorption spectrum; and due to the unique capability of RP chromatography to provide a structural recognition mechanism, they will also show different retention times. This is illustrated in Figure 2.5 which shows a reaction control analysis by LC-UV-MS to monitor the progress and yield for an N-aryl coupling reaction; unfortunately, the educt is not pure but is contaminated with a certain amount of the competing isomer – which could potentially lead also to an *E/Z* mix of product

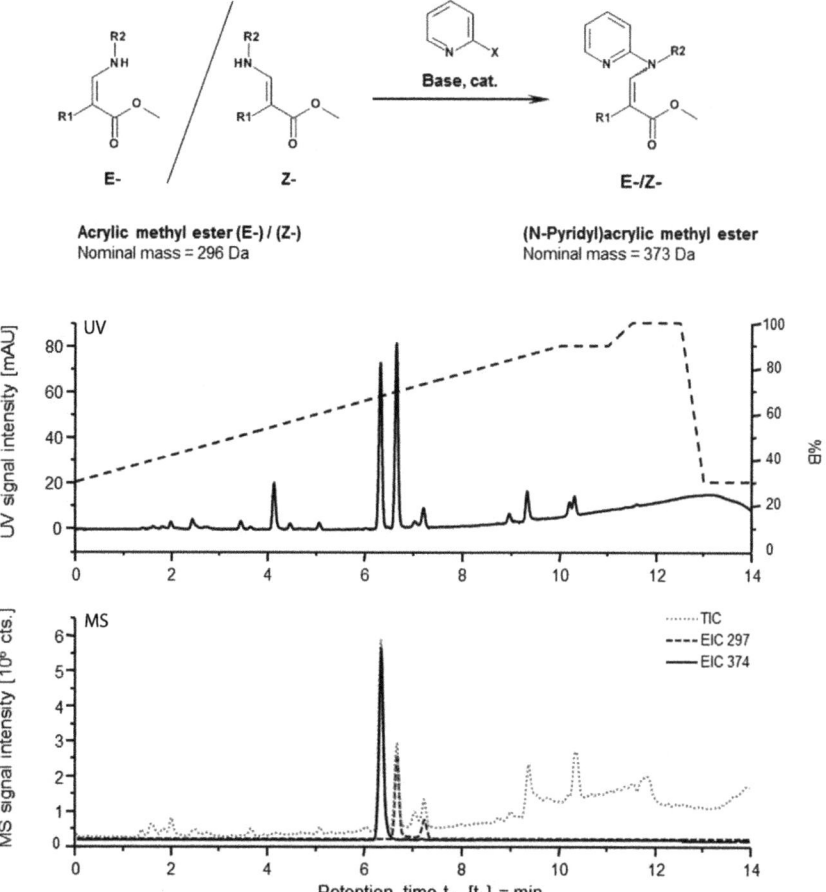

Figure 2.5 Distinction of isobaric compounds by RP chromatography and UV detection for a reaction control analysis (N-arylation of an E/Z acrylic ester mixture.)

molecules. The coupling reaction indeed is stereoselective as expected (only one product isomer can be detected); however, based on the LC/MS chromatograms alone, there is no way to correctly assign the two educt peaks to the related E and Z isomer – the extracted ion chromatogram (EIC) for the educt mass just shows two peaks belonging to the two educt species, without any further indication of which one is which. However, as the E and Z educt isomers show significantly different UV spectra (not shown) and RP chromatography provides you with an excellent separation of the two species, peaks can be assigned correctly based on the diode array detection spectra.

Whenever you plan to combine a mass spectrometer with a second detector, it is this additional detection principle that tells you how to technically realize the hyphenation. Don't forget – the mass spectrometer always destroys your analyte while measuring and detecting it, so it must be the last detector in your instrument arrangement. If you want to add a nondestructive detector to your system – all spectroscopic detectors belong to that – you can simply connect this in line with your LC column upfront and the mass spectrometer behind. One thing to take care of then is the additional volume of the detector flow cell which in most cases also adds a measurable contribution to band broadening. Using UV detectors, you would thus ideally look for a flow cell with a very small internal volume at maximum light path length. Typical UV flow cells of conventional design are commercially available with volumes down to 2–5 µL. A very popular topic these days are flow cell designs featuring a capillary-based flow cuvette, glass-fiber optics, and internal total reflection, enabling very long light paths of 10 mm up to 60 mm, combined with very low internal volumes, which allow for remarkably low levels of detection and a superb sensitivity. Unfortunately, these flow cell designs have two inherent drawbacks: next to their higher price, the term "high sensitivity" also applies to their lower mechanical ruggedness. Pretty much all these high-sensitivity cells have a pressure limitation significantly below 1450 psi/100 bar, which is considerably lower than for conventional UV flow cells. As long as the UV detector is the last in line, this does not really matter, but in combination with a mass spectrometer (or, for instance, a fraction collector) it does. Another alternative in this case would be specifically optimized capillary flow cells. They are made from fused-silica capillaries and can be used and handled like any other conventional flow cell; they combine excellent pressure resistance (up to 4350 psi/300 bar) with impressing low internal volumes of less than 50 nL. However, these cell types are highly limited in their linear range, which does not recommend them for quantitative analyses, but makes them the ideal UV monitor cell with no measurable band-broadening impact on a classical UHPLC separation on 2.1 mm or 1 mm ID columns.

By contrast, destructive detection principles like nebulizer-based detectors (CAD, ELSD, etc.) must be connected parallel to the mass spectrometer using a tee piece. This may sound clumsy, but it is not totally of disadvantage. Depending on the LC method settings, a postcolumn split may be a good idea anyway – it would allow you to run your LC separation with the best-suited linear velocity (in the van Deemter minimum of your column or beyond) and still to reduce the flow rate entering the MS ion source. A potential drawback of the split flow approach would be the addition of another extra-column volume, potentially affecting your chromatographic efficiency and resolution. But in turn you would gain a very versatile and powerful analysis tool: combining a uniform detection principle like CAD with mass spectrometry accelerates screening experiments massively. One single chromatographic run in such a setup will give you a very good (semi)quantitative result from one of the most unspecific detectors commercially available, while a parallel-running HR/AM mass spectrometer gives you excellent qualitative data for virtually unambiguous compound identification. A very interesting extension in this context can be offered by electrochemical detectors (ECD). By making use of the redox activity of analytes to generate a detector signal, they will alter the analyte species; a MS in series behind will, thus, no longer detect the molecular ion mass of the initial species, but of the oxidized or reduced one. If you needed the unchanged mass of the original analyte, you would then add the ECD in a parallel split. By plugging the ECD in line with the mass spectrometer, however, you will get the highly interesting option to enhance the detection sensitivity of your MS by electrochemically converting your analytes into oxidized or reduced species which may show a much better signal response than the original molecule. You can even use this approach for bioanalytical applications, for instance by electrochemically mimicking certain metabolism processes and to investigate them immediately, on the fly so to speak, in the mass spectrometer.

Take-home messages

- Mass spectrometers are no universal detectors
- Mass spectrometers detect highly selectively, depending on their operation settings, but not specifically
- For the unambiguous compound structure confirmation, you will always need (at least) one additional structure elucidation method (e.g., NMR spectroscopy) in addition to your MS(/MS) findings
- In all cases where molecules cannot be distinguished based on their molecular mass within the experimental error (isobaric compounds), additional and complementary detection principles are imperative. The related detector modules can be added either serially or in parallel by splitting the LC column effluent, depending on the detection principle.

2.2
LC/MS Method Development and HPLC Method Adaptation – How to Make My LC Fit for MS?

When liquid chromatography meets mass spectrometry, two very different worlds with highly contrary physical requirements need to come together. While an HPLC separation works against atmospheric pressure at the outlet end, a mass spectrometer always asks for a high-quality vacuum to operate. The interface between LC and MS, the ion source, must therefore handle multiple tasks simultaneously: transferring the dissolved analytes into the gas phase, separating the analytes from the residual mobile phase (typically done by gas phase transfer), controlled ionization of the analyte molecules, and a focused analyte ion transfer into the evacuated mass analyzer. None of these jobs is a simple one – just take the removal of the mobile phase: as we know, the molar volume of a gaseous compound is 22.4 L under normal and 24.5 L under standard conditions. Hence, water at a flow rate of 1 mL/min, equaling 1/18 mol/min, forms 1.2 L of vapor every minute, which needs to be completely removed from the analytes and drained out of the ion source. Therefore, the LC separation needs to meet certain limiting requirements to ensure a smooth signal generation in the mass spectrometer.

2.2.1
Method Development LC/MS – LC Fits the MS Purposes

The general approach to a new LC/MS separation method is not so much different from the way you would do it in classical HPLC analysis – it typically starts with the offline development of a new UHPLC method under the framework dictated by mass spectrometry requirements. In parallel, the mass spectrometer needs to be optimized to the analytes of interest in all its settings for the ion source, the ion transfer optics, and the individual mass analyzers. In the next steps, the result of both optimizations are then merged into the hyphenated UHPLC-MS method; total system functionality needs to be verified and matrix effects have to be determined before the whole method then undergoes the regular method validation. In a nutshell, this recipe is summarized like this:

a) Select the most appropriate ionization principle
b) Develop the LC separation offline
c) Optimize the mass spectrometric parameters offline
d) Make the connection while verifying the MS settings and determining matrix effects
e) Finally test the whole setup and validate the new method.

The individual steps on the way to the final method will be discussed in more detail in the following sections.

2.2.1.1 LC Flow Rate and Principle of Ion Formation

Let me open this section with a short remark on the term "sensitivity". In the first instance, "sensitivity" is defined as the slope of the response function which describes the signal change depending on a change in analyte amount or concentration. The slope, also known as response factor, always has a physical dimension (signal unity per concentration measure), which makes it impossible to directly compare response factors of different detection principles. However, the sensitivity is always linked to the ratio of signal intensity to baseline noise – a dimensionless measure which is also used to determine limits of detection (LoD) and quantitation (LoQ) (and which allows a comparison between different detection methods). The following discussion will always cover both of these interpretations.

It is already the selected ionization principle that tells you about the maximum LC flow rate you should confront your mass spectrometer with. A comprehensive overview on all known and commercialized ionization techniques is given in Chapter 1. The far majority of all LC/MS applications, however, is served by either one of only two principles, Electrospray ionization (ESI) or atmospheric pressure chemical ionization (APCI).

Electrospray ionization (ESI) which is applied in about 82% of all published online-LC/MS hyphenations (the rest is shared between APCI with 16% and APPI and others with 2%; LC-MALDI-MS in contrast is a classical offline hyphenation example [6]) enables the use of 50–300 µL/min flow rates at best sensitivity if assisted by a pneumatic nebulizer gas [8]. Pretty much all commercial ESI interfaces (excluding nanospray sources) can be operated at much higher flow rates of up to 1 mL/min and more. While being a concentration-sensitive process, ESI by theory is barely affected by the flow rate; the peak height should not change significantly with the LC flow rate. Indeed many literature examples demonstrate that the sensitivity of ESI methods suffers at elevated flow; however, this is only a sign that the excess of mobile phase cannot be removed effectively anymore. This can happen at LC flow rates beyond 1 mL/min, depending on the ESI interface design and the efficiency of the source heating or the supporting nebulizer [9]. In reality, experimental conditions like composition and mixing change of the mobile phase over time can lower the sensitivity already at flow rates higher than 300–500 µL/min. It is hard to predict the extent of this reduction as there is no mathematical model or a rule of thumb for this; so it is highly recommendable to monitor the signal intensity and the signal/noise ratio for your target analytes at different flow rates by a flow injection analysis. Depending on the solvent removal capacity of the ESI interface, the maximum sensitivity might be reached at flow rates which are lower than the van Deemter minimum of the LC column packing material. Then it is a case-by-case decision where to set the priority, on highest detection sensitivity or best chromatographic efficiency.

Atmospheric pressure chemical ionization (APCI) tolerates much higher flow rates than ESI, which lies in the nature of the process – it simply needs a minimum amount of solvent vapor to create the reagent gas that is responsible for the analyte ionization. APCI is a mass-sensitive process [10], which benefits from higher flow rates because more analyte molecules per time enter the APCI interface, thus, leading to increased peak heights at higher flow rates (the literature, however, also discusses deviations from this general rule [11]). The operation range of APCI starts at 150–200 µL/min and ends at maximum 2 mL/min, with a sensitivity loss being observed also here at very high flow rates, depending on the interface design and the evaporation capacity. Just like ESI, monitoring the sensitivity in dependency of the flow rate should always be done to determine the ideal LC/MS flow rate during the MS method development. Table 2.4 summarizes the usable and the most effective working ranges of ESI and APCI.

Hence, APCI seems to be a very suitable interface principle, particularly for fast UHPLC-MS separations. It should also be mentioned that APCI is less prone to matrix effects in many applications, thus, having the tendency of being more ro-

Table 2.4 Applicable and ideal working ranges for selected ionization processes.

	Applicable working range	Ideal working range
Nano-ESI source (without nebulizer gas)	< 5 µL/min	20–800 nL/min
Standard ESI source (with nebulizing support)	0.01–1.5 mL/min	0.05–0.3 mL/min
APCI source	0.2–2 mL/min	0.3–1 mL/min

bust and accurate than ESI (refer also to Section 2.3.4.1). However, the user is not free in his or her choice, as the analyte properties dictate which ionization principle has to be applied. Due to the analyte polarity, ESI is the first choice in most cases. However, a dedicated LC column hardware can comply very well with the low flow rates ideal for ESI: columns of 2.1 or 1 mm inner diameter allow operating even UHPLC stationary phase materials with average particle diameters below 2 µm at optimum linear velocities, as this translates in still very low volume flows in the µL/min range at those small column inner diameters. If the sensitivity loss with ESI at ideal chromatographic linear velocity was still too high, a postcolumn split of the LC effluent could be a good way out of this dilemma. Such a split is easily realized by a tee piece and two restriction capillaries – their dimensions will determine the split ratio between the primary flow to the MS and the bypass to the waste. As an extension, the split bypass does not mandatorily have to go to the waste: it can also be used for a second, ideally mass-sensitive detector, as already discussed in Section 2.1.4. Nevertheless, such a split always bears also the risk for band-broadening void volumes, so great care must be taken while selecting the different pieces and assembling the split construction.

2.2.1.2 The LC in LC/MS: MS-compatible Phase Systems, Eluent Compositions, and Additives

A MS-compatible chromatographic separation needs to meet certain requirements dictated by mass spectrometry if it should run seamlessly. The whole story begins with the right selection of a MS-compatible phase system, so the choice of an appropriate combination of stationary and mobile phase.

As in LC standalone, the predominant retention mechanism for LC/MS applications is reversed-phase (RP) chromatography. In theory, practically all common RP phases are suitable for both LC and LC/MS separations. Modern separation columns have already been developed with LC/MS compatibility in mind, thus, providing stable bonding and low metal content; older phase materials, however, can sometimes show a pronounced bleeding (i.e., a leaching of the bonding) which leads to an enhanced baseline noise in the LC/MS chromatogram. Typical column inner diameters for LC/MS applications should not exceed 2.1 mm. This ensures that the LC separation can be run (at least) in the minimum of the van Deemter curve of the given phase material, resulting in the best chromatographic efficiency, without overloading the MS ion source with too high flow rates. An interesting alternative to RP-based analyses is the separation by hydrophilic liquid interaction chromatography (HILIC), particularly for very polar analytes. This mechanism works with elution gradients from a high to low percentage of organic modifier; as a direct consequence, highly polar analytes elute at fairly high organic content in the mobile phase, which significantly improves the signal intensity in the mass spectrometer, as discussed in the next section. However, the number of analytical challenges that can be solved by HILIC is rather limited, and HILIC can also have certain additional implications like long equilibration times. Therefore, RP chromatography will certainly continue to be the first choice for new LC/MS applications.

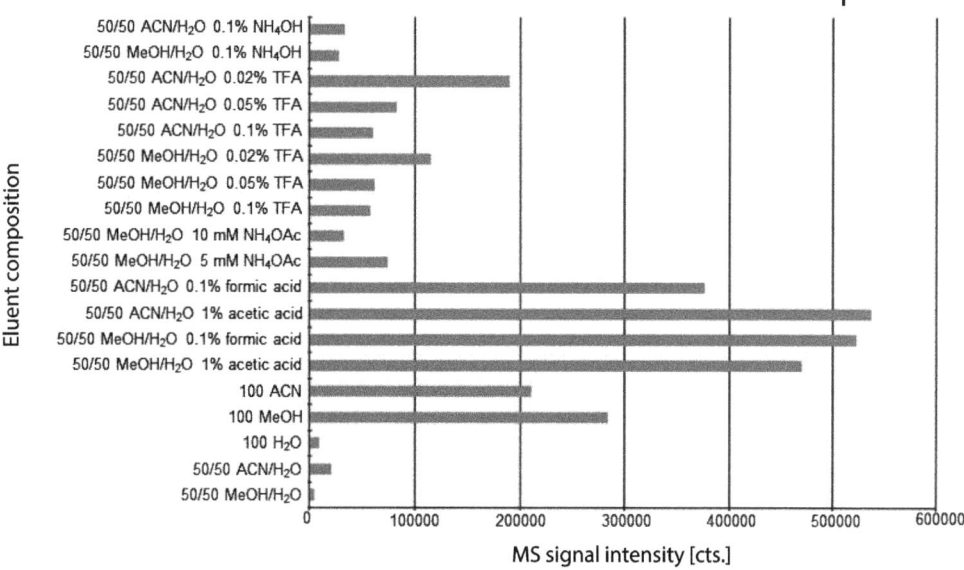

Figure 2.6 Signal intensity of Leu-enkephalin dissolved in various common LC/MS solvents; data acquired in ESI(+) mode during syringe pump infusion [12].

MS compatibility of the mobile phase means that *all* ingredients of the eluent have to be volatile. Nearly all solvents used in RP chromatography comply with this rule; water, which is the chromatography liquid with the highest evaporation enthalpy, is very well compatible with ESI and APCI processes – even more, a minimum amount of water is vital for an acceptable ionization yield. Organic solvents enhance the spray drying not only due to their higher vapor pressure but also by reducing the surface tension of the solvent droplets in the electrospray, which facilitates the evaporation of residual solvent molecules. As a practical consequence, higher organic content leads to a better spray stability and increased signal/noise ratio – which can also be observed in any RP gradient run; in electrospray ionization, the signal intensity typically increases linearly with the increase in organic content in the mobile phase at up to 80% [9]. The signal intensity in the mass spectrometer measured in mixtures of solvents hereby can significantly deviate from those in pure solvents. Figure 2.6 illustrates the ion yield, reflected by the MS signal intensity in counts, of the pentapeptide Leu-enkephalin which was dissolved in a variety of MS-compatible solvents, infused into an ESI source by a syringe pump, and being recorded in positive mode [12]. For a more in-depth discussion the reader is referred to the literature [13]. At the end of the day, the actual solvent composition during the analyte ionization is dictated by the eluent condition of the UHPLC method at the time of elution; the user has therefore only a limited chance to vaporize and ionize the analyte under the most suitable conditions.

Volatility, however, must also be a given for all kinds of additives to the mobile phase. All modifiers forming nonvolatile salts or precipitates lead to a massively enhanced suppression of ion formation in the ion source (*ion suppression*) and to a rapid and significant contamination of the ion source with salt crusts. In addition to the frequent cleaning efforts, this leads inevitably to a pronounced sensitivity loss in your MS chromatogram, although many major instrument vendors claim the opposite. As a consequence, most of the classical buffering agents, acids, bases, and additives well-known in LC standalone like phosphate or borate buffers, and more generally alkali and alkaline earth metal salts have to be avoided in combination with LC/MS. Instead, various (semi)volatile organic acids, bases and their respective ammonium salts are viable alternatives. For acidic pH, formic acid, acetic acid, and trifluoroacetic acid are most popular, while aqueous ammonia solutions or alkyl amines, for example, triethylamine, cover the alkaline range. If the buffer capacity is needed, ammonium salts like ammonium formate, acetate, or bicarbonate are of first choice. Unfortunately, the neutral pH range around 7 is not well covered by LC/MS-compatible additives, while the non-MS-compatible phosphate salts have a strong buffer capacity around their second acidic constant with a pK_a of 7. All buffers based on the ammonia/ammonium equilibrium work best at a pH around the pK_a (NH_3/NH_4^+(aq)) of 9.25, whereas the counter ions from organic acids buffer well at pH of 5 or less (for instance, pK_a (acetic acid/acetate) = 4.75). An ammonium acetate salt solution dissolves in water to give a pH of approximately 6.7, which is exactly between the buffering regimes of the ammonium and the acetate system; without pH adjustment, an ammonium acetate solution is no buffer per se. Oxidizing agents are also not appropriate for LC/MS eluents: it is known that chloride ions not only promote ion suppression, but the electrospray can oxidize them into chlorine, which chemically modifies the analyte and over time also attacks your ion source hardware like the spray needle [14]. Volatile ionic detergents are also not a good choice, as they can deposit on the surfaces of the ion optics and the mass analyzer when entering the mass spectrometer; this can lead to electrical discharges and instrument malfunctions over time.

In an ideal world, it is highly recommended to not use MS-incompatible mobile phases in an LC/MS system ever in its lifetime. Yes, it is principally possible to flush a system which was running on a phosphate buffer as long as it takes to get it free from phosphate and alkali metals. However, it can take days or even weeks until the last remainders of ionogenic residues are washed out of the system and the column so that you will not see them disturb the mass spectra anymore. The simplest way to avoid this is dedicating LC/MS strictly to LC/MS compatible conditions only and not to mix up LC standalone conditions on LC and LC/MS systems.

But it is not only the type of eluent additive but also the concentration which matters. LC methods typically benefit from a high ionic strength of the mobile phase, as this provides a good pH buffer capacity or an effective suppression of secondary electrostatic interactions between polar or charged analytes and active sites of the stationary phase. However, this comes along with a high conductivity which negatively affects the ionization process in the ion source. Practically, the additive concentration should not exceed 50 mmol/L, with singly charged additive ions being in favor to multiply charged ones.

A topic frequently and controversially discussed in the literature should briefly be mentioned here, that is, the pros and cons of using formic (FA) or trifluoroacetic acid (TFA) as a modifier in LC/MS applications. These organic acids generate a pH of less than 2 up to 3, depending on the concentration, while still being MS-compatible as they do not form any nonvolatile precipitates in the ion source. Simultaneously, their anions – formate and triflate – are also ion pairing reagents, thus, actively controlling the retention mechanism of many analytes. Hence, cationic analytes potentially show a higher retention on a RP stationary phase. In many cases, an improved peak shape with lower asymmetry can also be observed. However, this effect, which is much appreciated in liquid chromatography, has its drawbacks in the MS ion source, as it compromises or even suppresses the formation of ions in the gas phase – a classical conflict of interests for the LC/MS analyst. Trifluoroacetic acid hereby is a much stronger ion pairing reagent than formic acid and interacts much stronger with RP materials. This leads to a better chromatographic elution behavior with enhanced retention and reduced peak asymmetry, but it also comes at the price of a noticeably worse signal-to-noise ratio in the MS chromatogram due to stronger suppression of ion formation in the ion source. When analyzing small molecules, the retention-enhancing effect of TFA is not much pronounced, but you clearly will observe ion suppression. Therefore, formic acid is more appropriate for the LC/MS analysis of small molecules. A classical tradeoff is the analysis of larger biomolecules, like peptides or proteins, which need to be detected in very low concentrations. For best sensitivity, formic acid would also here be the modifier of choice, but TFA can significantly improve the chromatographic performance. For bioanalytical questions, which frequently are realized by including a trap column in an online enrichment setup, it is therefore recommendable to use TFA in the trap column flow path, but to change to FA in the analytical separation; in some cases, however, the addition of TFA to the analytical flow path may be inevitable. Figure 2.7 illustrates nicely the retention-enhancing effect of TFA.

Take-home messages

- Use a "MS-friendly" phase system and optimize the LC method by UV detection
- Use stationary phases with low bleeding
- Use volatile eluent systems, typically made of volatile organic acids or bases of less than 50 mmol/L in water/methanol or water/acetonitrile mixtures
- Avoid nonvolatile salts, reactive eluent additives, and all types of detergents
- Aim at a percentage of organic modifier at the point of elution of 10–20% or higher
- Ideal flow rate for ESI: 0.05–0.3 mL/min
- Ideal flow rate for APCI: 0.3–1 mL/min.

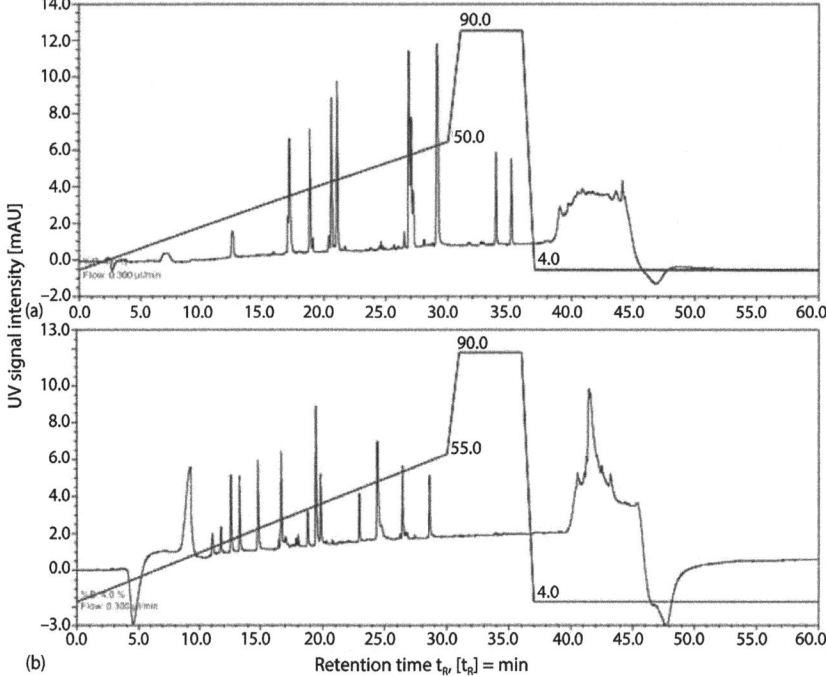

Figure 2.7 Separation of a cytochrome C digest by adding 0.05% TFA (a) or 0.1% FA (b) under identical chromatographic conditions

2.2.1.3 Optimizing the Mass Spectrometer Settings

Next to finding the right settings to run your UHPLC separation, optimizing the mass spectrometer parameters toward the analytes of interest is the second optimization step in the LC/MS method development. This *tuning* of a mass spec should not be mixed up with the *calibration*, where well-defined reference standards with exactly known molecular mass and fragmentation pattern are used to calibrate the mass analyzer(s) and the ion detector of the mass spectrometer. To tune the MS, a solution of the target analytes is continuously infused into the ion source by a syringe pump. In this flow-injection analysis (FIA), all relevant ion source and ion transfer settings can be optimized conveniently. It has been found appropriate to prepare the tuning solution at a target analyte concentration which is about 1/10th of the expected concentration in the real sample. The infusion solvent should ideally reflect the eluent composition of the sample zone at the point of the chromatographic elution. If this cannot be determined or approximated, you can also use the mobile phase composition averaged across the gradient program; however, in this case some deviation between the spray stability and sensitivity of the tuning and the later LC/MS separation has to be expected. The flow rate for the tuning experiment must equal the flow applied in the LC separation. If this is too high to be delivered by the syringe pump, then the UHPLC system needs to support here. The LC pump then provides the main flow rate for the mobile phase, while the tuning solution is then fed by the syringe pump via a tee piece. In this case, the tuning solution would be prepared as a concentrate, so that the final dilution at the target level is achieved by the flow rate ratio between the LC and the syringe pump; vice versa, the total flow rate contributions are distributed between LC and syringe pump according to the dilution ratio required for the tuning concentrate. As an example, imagine you need to tune the mass spectrometer for a LC separation flow rate of 500 µL/min; this would be pretty difficult with a syringe pump, as even with a 2 mL syringe it would give you only 4 min of time or less to tune your mass spectrometer before you need to refill. Making up a stock solution for your tuning run of ten times the target concentration which is then spiked into the UHPLC flow stream via tee piece allows you to work with 50 µL/min for the syringe pump, along with 450 µL/min for the mobile phase by the UHPLC system. Even with a very common syringe size of 500 µL, you will have about 10 min for the mass spectrometer tuning before the need for refill, which is much more convenient than the original plan.

While the infusion is running, the individual parameters for the ion source are optimized first, stepwise and one after another, monitoring the result closely in the online plot window of the mass spectrometry software unless your software offers a fully automated tune routine. Primary goal is to achieve a stable, continuous baseline without steps, breaks, and positive or negative spikes; a smooth and constant baseline is the perfect indicator for a stable spray and ionization process. In the case of nano-LC sources and some analytical-flow ESI sources, it is additionally possible to play around with the geometrical arrangement of the spray unit relative to the MS vacuum orifice and also the protrusion of the spray needle off the spray unit assembly. From experience, this matters substantially for the signal quality and intensity in nano-ESI sources by enhanced ion yield and ion transfer into the high-vacuum part of the mass spectrometer; for analytical-scale separations with several dozens or hundreds of microliters per minute of flow rate, the factory default settings for the spray assembly do not leave a lot of room for further improvement. The highest impact on spray and baseline stability have the gas flow and pressure settings of the source for pneumatically-assisted ESI and APCI as well as the drying temperature. In return, higher signal intensity is realized by an enhanced (selective) ion yield and an improved transfer of ions from the ion source into the ion transfer section in the vacuum part of the mass spectrometer. These factors are primarily addressed by the high voltage for the ionization process and, with APCI only, by the charge density at the corona needle tip, as well as via the acceleration voltages along the ion transfer path and the ion optics settings like the values for skimmer electrodes and focusing quadrupoles (for in-source decay, see Section 2.3.4.3). Unfortunately, these settings are not all independent from each other. Finding a global optimum for best sensitivity can be of some complexity, which is not always feasible under the time pressure of the daily lab work. However, following the sequence of optimization steps as described here commonly leads to a stable, practical and reliable working space.

Once the tuning has been finalized and the optimized MS detection parameters are set, it is recommended to record a continuous reference dataset for

1–5 min under the FIA conditions. This record is then evaluated with respect to spray/baseline stability and baseline noise in the LC/MS data trace, as well as the spectrum quality and purity, signal-to-noise for the relevant isotope signals, the isotope intensity pattern, and the obtained resolving power for an averaged set of mass spectra. When looking at HR/AM experiments, also the mass accuracy as the difference between the theoretically expected and the practically determined mass in ppm should be checked.

Take-home messages

- Dissolve the sample in the LC solvent, ideally at 1/10th of the expected concentration in the injected sample
- Infuse this solution into the mass spectrometer at the flow rate of the LC separation; if a syringe pump is too slow, use the LC pump for mobile phase transport and infuse a sample concentrate via a tee piece
- Adjust gas flows and dry temperature to yield a stable spray (typically not required in routine: ESI spray capillary adjustment)
- Optimize the settings for ion source and ion optics
- Record a reference dataset for 1–5 min
- Evaluate signal stability, quality/purity, signal-to-noise, and resolving power of the mass spectra obtained.

2.2.1.4 Make the Connection – Verifying Mass Spectrometric Settings and Determining Matrix Effects

The concluding step now deals with merging LC separation and MS detection. The user in a hurry will just install the separation column, make the fluidic connection between the UHPLC system outlet and the mass spectrometer inlet, and start the separation of a (matrix-free) standard solution under the conditions separately optimized for UHPLC and MS. There is already a pretty good chance that this approach is running straightaway. However, in many cases some fine-tuning and corrections might be inevitable. A very obvious reason is the separation running in gradient mode, where starting and end conditions of the mobile phase can differ significantly from the composition of the tuning runs. The tuning conditions optimized for the isocratic, averaged mobile phase might be more or less inappropriate for very different portions of water or organic modifier, which holds particularly true for the water-rich, "wet" part of the gradient program. In extreme cases, it will not be possible to find one common set of parameters which enables a stable baseline and detection throughout the whole gradient. It will be necessary to run different segments of the analysis with different ion source settings. Most commercial MS control software packages do not allow a continuous change of source settings over time to update for instance the nebulizer pressure with better-matching values across a gradient separation; what typically works as a compromise is to divide the elution window into different time segments for each of which different MS parameter settings can be defined. Fortunately, the majority of LC/MS applications work reliably with one global set of parameter values and hence do not need such a sliced approach.

Analysts who prefer the more thorough, in-depth approach may want to take an additional step prior to the on-column injection of the sample: to do so, the UHPLC-MS system with the separation column installed is connected to the MS, but similar to the tuning step, the sample is not injected by the autosampler and running through the column, but it is continuously infused postcolumn via syringe pump and tee piece. While running the separation gradient program by the UHPLC system, it can be immediately checked if the MS settings from the tuning run apply to the whole gradient composition, how the analyte signal varies with the solvent composition change in the mass spectrometer, and how the signal for the blind gradient changes due to solvent impurities, increased column bleeding, or compounds leaching from the system (see also Section 2.3.4.4). With the same setup, matrix effects like ion suppression can also be determined very efficiently, as previously discussed in Chapter 1. After such a final proof of functionality, the new LC/MS method can be validated following the same rules as for LC standalone applications.

Take-home messages

- The settings obtained during the MS tuning might not work well under gradient elution conditions of your LC separation due to deviating solvent composition
- Quick-shot approach to fix: set up the hyphenated method, inject a sample and modify the MS settings on the fly
- Thorough approach to fix: run a blind gradient over the LC column and infuse concentrated analyte solution postcolumn, and modify MS settings accordingly
- Determine matrix effects by postcolumn infusion experiments.

2.2.2
Converting Classical HPLC Methods into LC/MS

The more LC/MS methods continue to spread into routine analysis like pharma quality control (QC), the more LC methods developed from scratch are designed already considering the requirements from mass spectrometry; method developers would therefore be well-advised to envisage MS compatibility during method development even if there is currently no mass spectrometer planned in their lab. But how is it with traditional LC methods, if you're asked to bring them to the mass spec? A common case looks like this: a QC analysis for a drug substance based on phosphate buffer reveals an unknown impurity. The plant leader now needs to have the impurity identified to take a decision on how to move on, which is a classic example for the involvement of LC/MS. There is no patent remedy that always reliably leads to a 100% success rate, but some considerations will certainly lead to good results. The simplest approach would be to inject the sample of interest into a routine LC/MS separation setup with a very generic gradient and any random kind of RP column. Looking at the large variety of different RP selectivities among the commercially available LC columns and the difference between the properties of the original phosphate-buffered eluent and the MS-compatible counterpart with, for example, formic acid, there is a pretty good chance that the resulting LC/MS chromatogram looks quite different from the LC original; this in return can hamper the unambiguous identification of the peak of interest. It is typically more promising to mimic the original LC separation as well as possible by working on a column of the same type as the original and by applying the same elution conditions, with modifications of the mobile phase toward better MS compatibility. Acidic phosphate buffers should be replaced by either formic acid or trifluoroacetic acid in the same pH range. For neutral to alkaline pH, ammonium salts should be used as they also have pH buffering properties even if the buffer points are not always matching those of the LC classics such as phosphate, borate, and others (see also Section 2.2.1.2). In case you'll try to rebuild the previously MS-incompatible LC method, it is a good idea to try this with a fresh column off the shelf which has never seen the original mobile phase before. Every stationary phase has some kind of a chemical memory which means that the properties of the stationary phase alter over time depending on the type of mobile phase used, the number and amount of samples injected, and the type of matrices. Particularly phosphate is known to permanently change the selectivity of some RP phases, which means that a LC column which has been previously run with phosphate eluents will very likely not show the same selectivity than a virgin column, even if both are run under phosphate-free conditions. In addition, a fresh column will not suffer from the permanent leaching of phosphate ions that can negatively affect the detection in the mass spectrometer.

2.3
Pitfalls and Error Sources – Sometimes Things Do Go Wrong

The previously discussed considerations cover very fundamental aspects of an LC/MS method setup; they are discussed typically once, at the beginning of your work, as they deal with generic questions like "which instrument should I use" or "how should my LC/MS method look like". Once decided, these things do not change across the lifetime of a method. But there are many minor or major deviations from the routine happening in the lab every day. Surprises and pitfalls are many, from wrong mass assignments or unknown mass signals, noisy MS spectra to a totally empty mass signal and sheer baseline noise. This final section should illustrate very common error symptoms of LC/MS lab life and to point out ways how to tackle them.

2.3.1
No Signal at All

The most striking problem while running a sample analysis is described very quickly: after thoroughly developing a robust and sensitive LC/MS analysis method, you start an injection and you see – nothing. Obviously, the root cause for this "nothing" can be either the LC or the MS part of your hyphenated LC/MS system, with the worst case being that both chromatography and mass spectrometry do not run as expected. The hard truth is that in most of these cases, the instrument hardware is operating properly – none of the diagnostic tools for the devices report any error for the electric and electronic instrument subassemblies. Most of the time, it is an inadequate ion formation which is responsible for weak or no MS signals, so the system struggles somewhere with the ion source, the ion transfer to the vacuum section of the MS, or with the sample introduction. Technical defects behind the ion source, be it in the ion optics, the mass analyzer, or the mass detector, are observed much less frequently.

To pin down the error source, it is very helpful to check first the signal intensity and the noise level of the baseline both in the LC/MS chromatogram and in the mass spectra via the online view of the instrument control software. A very low signal intensity of only several hundreds of ion counts and a noise pattern that looks much more like an erratic flaring of signal spikes than a continuous base level signal are strong indicators that virtually no ions at all are reaching the mass analyzer or the mass detector. The core reason for this is typically a quite rough mechanical or electrical defect either in the LC or the ion source sprayer assembly. To then further troubleshoot the LC part of your system, a UV detector is of invaluable help – no other detection principle is par with UV with respect to ruggedness. Nowadays, dedicated UV monitor flow cells of only a few dozens of nanoliters internal volume allow even for a permanent monitoring of your LC/MS separation all the time as they do not contribute measurably to band broadening and, thus, reduced peak efficiency. This allows for a rapid and doubtless verification of your chromatographic separation running properly. Having confirmed that chromatography separates your analytes and transports them toward the outlet of the LC/MS connection capillary accordingly, there must be something wrong within the MS ion source. For instance, a deformed or even broken spray capillary or needle prevents effectively the LC effluent from entering the MS ion source; hence, no stable nebulizing spray can be created. Very popular here is a spray needle tip which is cracked, resulting from a short touch on the lab bench or during a careless insertion into the source assembly. ESI and APCI ion sources that apply the high voltage (HV) for the electrospray or ion formation to the spray needle (like SCIEX and Thermo Scientific devices) instead of the MS vacuum inlet (Bruker Daltonics, Agilent Technologies) can additionally suffer from damage in the electrical wiring or plug connections due to improper handling.

With a chromatography and spray unit assembly being intact and a baseline noise that indicates a reasonably stable spray being formed, obviously your system generates ions properly, but for whatever reason too small of an amount of ions truly reaches the mass analyzer. The ion current measured along the ion transfer capillary in the evacuated part of your mass spectrometer is a good indicator for an acceptably high ion transfer yield. In the case of this being very low, your first step of action should be a thorough cleaning of the ion source including the ion transfer capillary. Analyte or dirt deposits on the various surfaces which you cannot even see with your eyes can suppress ion formation and need to be removed by cleaning. It is highly recommendable to check the state of the spray needle, particularly the tip, with a magnifier or a suitable microscope. If the needle tip is bent or cracked, you will need to replace it. Depending on the sprayer assembly design and thus MS vendor, the alignment and positioning of the spray needle in the sprayer assembly should also be checked; a re-adjustment of the needle alignment can be very helpful, as the extent of the tip protrusion out of the sprayer assembly can change the detection sensitivity by several orders of magnitude.

2.3.2
Inappropriate Ion Source Settings and Their Impact on the Chromatogram

The baseline quality of our LC/MS chromatogram is a good indicator if the processes within your ion source are running seamlessly; the same for the base signals in every individual mass spectrum. Typical error patterns can mostly be divided into two categories: increased baseline noise and poor baseline stability. Both phenomena suggest that something is going wrong with the selective creation of analyte ions or the continuous removal of residual solvent, for instance due to spray instability. An increased baseline noise is mostly a clear sign for dirt in the MS ion source, as it is the result of too many different ions being created over a wide *m/z* range and entering the mass spectrometer simultaneously. The reasons for this so-called *chemical noise* can be manifold. The first thing to check would be the cleanliness of the ion source and when it has been cleaned most recently. Any residues on the spray needle tip, on the internal surfaces of the ion source assembly, on the orifice and metal plates of the vacuum inlet, on the ion transfer capillary, or the first stages of ion lenses are very good candidates for increased baseline noise. Obvious sources for these chemical contaminations are not only sample or matrix components but also wanted or unwanted parts of your mobile phase, and the dry gas or nebulizer gas of the ion source. The LC solvents used to compose your mobile phase should therefore always be of LC/MS purity grade (labeled "LC-MS grade", "ULC/MS", etc., depending on the supplier). The very popular "gradient grade" purity solvents, however, are merely optimized for a minimal amount of UV-absorbing impurities and therefore are typically not pure enough for LC/MS applications.

The gas used to dry and/or to nebulize the LC effluent – typically nitrogen – can also introduce contaminations into the ion source and, thus, reduce your signal quality. The gas purity level should be at least 99.0% or better 99.5% and higher. Depending on the nitrogen origin, the types of contaminants can vary. Many mass spectrometers have a very high gas consumption of up to 10–25 L/min, especially when combined with analytical scale LC flow rates of several hundreds of µL/min. This locks out the nitrogen supply from gas cylinders – although this ensures very high gas purity, the average gas consumption of a mass spectrometer would typically empty these cylinders every 1–2 days, which is not very economic and not convenient either. Higher amounts of nitrogen can only reasonably be provided by nitrogen generators or a nitrogen supply line in your lab based on liquid nitrogen evaporation. The latter typically ensures the highest gas purity grade. Nitrogen generators – which misleadingly do not *generate* nitrogen but just purify air – are fed with pressurized ambient air and remove the oxygen by a membrane separator. The pressurized air, however, is provided by a compressor and therefore frequently contains residual oil mist and other hydrocarbons diffusing out of the compressor hardware. These contaminants must be effectively eliminated from the nitrogen stream by gas filters based on activated carbon filter assemblies or other adsorbents. If such a filter is lacking or the filter is fully loaded and needs to be replaced, you will also observe a significant rise in the baseline noise.

Next to these "real" contaminations originating from unintended chemicals, gas phase aggregates or *clusters* consisting of residual solvent molecules and charge carriers can be formed under nonoptimized ion source conditions. These clusters can in summary be heavy enough to create signals within the m/z detection range of your MS experiment and thus permanently contribute to the baseline noise. A thorough optimization of the ion source parameter settings can effectively suppress this cluster formation. Figure 2.8 illustrates this nicely by depicting the MS signal intensity of a target analyte (Astemizol, m/z 459.3) introduced by flow-injection analysis (FIA) under varying dry gas temperatures. In Figure 2.8a, we see that the individual analyte signal intensity in the extracted ion chromatogram (EIC) remains nearly constant with rising dry gas temperature; the analyte ion yield does not really change with higher temperatures. In Figure 2.8b, however, the overall noise level in the total ion chromatogram (TIC) significantly varies

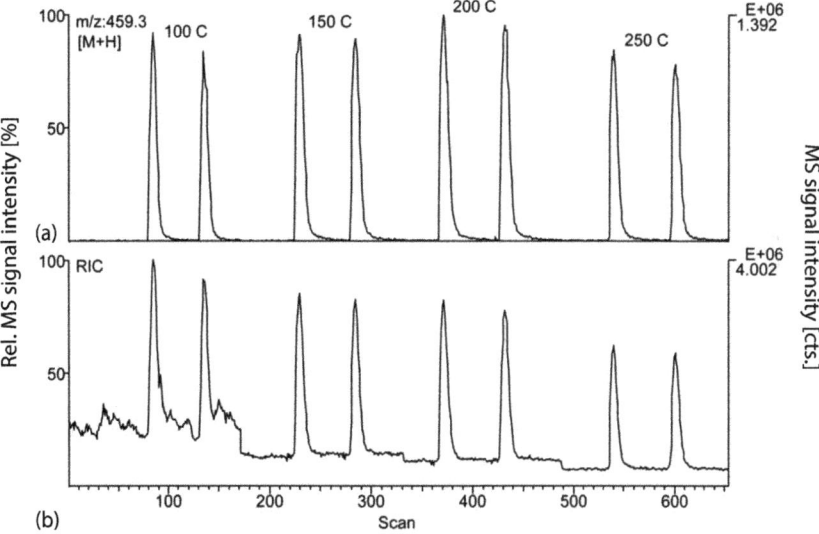

Figure 2.8 Influence of the dry gas temperature on MS signal quality using FIA of astemizol in 10 mM aqueous ammonium acetate/methanol 20/80 (v/v) at 50 μL/min on a triple quad instrument; (a) extracted ion chromatogram of astemizol ([M + H]$^+$), m/z 459.3; (b) reconstructed total ion chromatogram.

with the drying temperature, which is a clear sign for a more efficient gas phase aggregate destruction (*declustering*).

Finally, an increased baseline noise can also be caused by too high of an electrospray voltage (called ESI voltage in the following). In extreme cases, you can even see this with the naked eye by a pale blue glow discharge on the ESI needle tip. This glow discharge is facilitated by too many charged species in the mobile phase, for instance, by using a highly concentrated buffer salt, acid, or base; in addition to a disrupting ion beam in your MS source, this can even lead to a voltage flashover in parts of the ion optics. You should then try to reduce the ESI voltage, the buffer concentration in your eluent, or both.

Baseline stability issues, in particular spikes or spontaneous drops, typically indicate an improper nebulizing or an unstable electrospray. A poor nebulizer gas flow or pressure rate as well as a nonideal ESI voltage then lead to the formation of larger liquid droplets or clusters in the ion source; these droplets tearing down from the ESI needle will lead to negative drops in the baseline, while a droplet burst releases many ions from this droplet simultaneously and, thus, generates characteristic positive spikes. This can even happen if the sample solvent contains a strong ion pairing reagent: injecting a sample acidified with trifluoroacetic acid can lead to negative baseline spikes even minutes after the inert solvent plug has left the separation column.

2.3.3
Ion Suppression

Whenever possible, the LC/MS method development is done with high-purity reference compounds. Real samples, however, can host significant amounts of impurities and matrix components; those can lead to a substantial signal intensity reduction for your analytes at the same identical concentration as in your purified method development standard solutions. This is a very frequent problem in matrix-rich sample analyses like food, blood and plasma, or cell tissue. The origin of this deviation in signal intensity is the co-elution of the target analyte with other compounds which interfere with each other during the ionization process and, thus, affect the ion yield for your analyte of interest. Depending on the mechanisms behind this, either a signal enhancement or a signal reduction could be observed, the latter being called *ion suppression*, which is already discussed in Chapter 1. The reasons for a signal-enhancing or signal-reducing effect are manifold. An in-depth discussion would exceed the scope of this section, so the reader is referred to the literature [15]. Signal suppression can be observed both with ESI and in APCI interfaces, for positive as well as negative polarity. However, there are many examples indicating that APCI is less prone to ion suppression than ESI, and negative polarity seems to be less affected than positive. We already learned about how to qualitatively determine the signal enhancement or suppression effect using a *post-column infusion* setup (refer to Chapter 1). A quantitative assessment of matrix effects requires a comparative analysis of a spiked, matrix-free sample and a matrix-containing sample. A comparison of peak areas from both experiments will tell you then about the extent of a signal change by matrix effects. Unfortunately, matrix effects cannot always be eliminated, but at least they must be quantified so that they can be considered in real sample analyses. Ways to address them could be sample dilution or selective matrix depletion by dedicated sample preparation. As a concluding remark, it is explicitly this signal intensity change caused by matrix effects that passionately pleas for high-quality LC sample separation as an essential requirement for robust and reliable quantitation with mass spectrometry. Infusing an unknown and nondiscriminated bulk solution of your sample into a mass spectrometer will always have the potential for many disturbing interactions of the analytes with each other during the ion formation process; as a result, you would barely be able to assess if you truly see every analyte in your sample, not to speak of a reliable analyte quantitation.

2.3.4
Unknown Mass Signals in the Mass Spectrum

Many LC/MS users are permanently facing the challenge that they cannot plausibly assign all signals in a mass spectrum to a given target analyte. Either the measured mass signals differ from what is expected, or the spectrum shows more m/z values than predicted. In the following, a selection of root causes being responsible for this mismatch between experiment and expectation are discussed. However, it would be beyond the scope of this book to talk comprehensively about virtually all aspects of observing unknown mass signals; therefore the reader is also referred to MS-specific literature for further reading [7, 17, 20].

2.3.4.1 Gas Phase Adducts
The most frequent reason by far for unknown mass spectrum peaks is the creation of adducts between the analyte molecular ion and other low-molecular-weight ions and/or neutral chemical entities in the gas phase. Hereby, the type and extent of adduct formation varies substantially with the ionization principle, the ion source parameter settings, the analyte properties, and the quality of the mobile phase. However, gas phase adduct formation or *clustering* is not disadvantageous by default; it is even imperative for the conversion of a neutral molecular species into a charged one with ESI and APCI, as adding a proton to (in positive polarity mode) or subtracting a proton from (in negative polarity) an analyte of the molecular mass M converts it into the charged state of $[M + H]^+$ and $[M - H]^-$, respectively. Consequently, the measured m/z ratio differs from the theoretical one of the neutral species by the amount of one proton mass. But in addition to this fundamental prerequisite for the mass detection, many other charged adduct species can be observed in reality. The virtually ubiquitous sodium and potassium cations leaching from the glass surfaces of the solvent bottles, for instance, frequently lead to the respective adducts $[M + Na]^+$ and $[M + K]^+$. The longer the shelf life of your solvents, the more these sodium and potassium clusters are even favored compared to the proton adduct; therefore, shifts in the adduct ratio between proton and alkali metal-based ionic species can even be used as a rough estimate of

Table 2.5 Common gas phase adducts at positive (left) and negative (right) polarity. Mass differences refer to the difference between [M + H]$^+$ (left) or [M − H]$^-$ (right), respectively, and the related gas phase adduct.

Positive polarity Gas phase adduct	Nominal mass difference (Δ Da)	Negative polarity Gas phase adduct	Nominal mass difference (Δ Da)
[M + NH$_4$]$^+$	+17	[M − H + H$_2$O]$^-$	+18
[M + H$_2$O + H]$^+$	+18	[M − H + CH$_3$OH]$^-$	+32
[M + Na]$^+$	+22	[M + Cl]$^-$	+36
[M + CH$_3$OH + H]$^+$	+32	[M − H + CH$_3$CN]$^+$	+41
[M + K]$^+$	+38	[M + HCOO]$^-$	+46
[M + CH$_3$CN + H]$^+$	+41	[M + CH$_3$COO]$^-$	+60
[M + H$_2$O + CH$_3$OH + H]$^+$	+50	[M + Br]$^-$	+80
[M + CH$_3$CN + Na]$^+$	+63	[M + HSO$_4$]$^-$	+98
[2M + H]$^+$	−	[M + H$_2$PO$_4$]$^-$	+98
[2M + Na]$^+$	−	[M + CF$_3$COO]$^-$	+114
[2M + K]$^+$	−	[2M − H]$^-$	−

the solvent shelf life in your LC/MS system. With the sodium or potassium adduct becoming the most prominent *m/z* signal in the spectrum over time, it is high time to prepare a fresh lot of mobile phase for the UHPLC separation. Due to the well-known chemical similarity between the alkali metal ions and the ammonium ion, the use of ammonium salts as buffering agents in LC/MS applications will lead to the analog creation of the ammonium adduct [M + NH$_4$]$^+$ instead. Next to those species, also higher aggregates involving three or more − mostly neutral − molecules can be observed; depending on the solvent evaporation efficiency, excessive solvent molecules can then cluster with the analyte and the charge carrier, resulting, for example, in [M + H$_2$O + H]$^+$ or [M + CH$_3$CN + H]$^+$. Even clusters of multiple analyte molecules sharing one proton or sodium cation as [2M + H]$^+$ or [2M + Na]$^+$ can often be detected. All these adducts reveal very characteristic mass differences in comparison to the simple proton adduct. Table 2.5 lists the most common gas phase adducts with their respective nominal mass difference to the singly protonated ([M + H]$^+$) or deprotonated ([M − H]$^-$) reference. More in-depth information on that matter can also be found in the references [16].

In general, the trend to form those adducts can be controlled quite effectively via the ion source parameters. An appropriate setting for the drying conditions of the source (e.g., dry gas temperature, nebulizer gas pressure) allows the declustering of higher aggregates into less complex ones. Also, APCI is typically less prone to higher aggregate formation with, for example, alkali metals, as the charge transfer to the analyte molecule happens after the evaporation, that is, entirely in the gas phase. In ESI, in contrast, the charge transfer takes place parallel to the gas phase transfer, so while the analyte is still partially in the liquid phase. For the same reason, APCI is in many cases also less affected by matrix effects, depending of course on the application. As the cluster formation is also influenced by type and amount of mobile phase, the signal intensity of different gas phase species can change with the UHPLC flow rate.

2.3.4.2 Chemical Modifications of Analytes by the Separation Conditions

It is not always a simple gas phase adduct formation which is responsible for an observed difference between the expected and the measured analyte signal. In some – rather rare – cases, a true chemical reaction between the analyte and some components of the separation system can also be the reason for a change in the nature of the analyte. Two known phenomena are to be mentioned here. An *electrochemical oxidation in the ESI ion source* is typically caused by the presence of potentially oxidizing agents in the mobile phase. Even if no obvious chemical reaction in free solution under ambient conditions takes place, it is the influence of the strong electric field of the electrospray process that can trigger the formation of a redox partner in the nebulizing spray which then initiates an oxidation of the target analyte. One example is the previously mentioned addition of chloride salts to the mobile phase (see Section 2.2.1.2). The high voltage of the ESI source generates small amounts of chlorine species in the gas phase which then oxidize for instance proteins at suitable sites in their primary sequence, leading to a m/z difference of +16 [14]. As described earlier, this is a valid reason to eliminate chlorides from the mobile phase, in addition to the formation of nonvolatile precipitates. Another observed phenomenon is a *photochemical oxidation* induced by high-sensitivity UV detectors. High-energy UV light of low wavelengths can excite the electrons of selected analytes with an appropriate UV absorption maximum so that this induces oxidation reactions. Large biomolecules are more prone to this effect than small organic molecules. In case of doubt, there is a very simple way to verify this suspicion by just repeating the LC/MS experiment without the UV detector being active between column and mass spectrometer. The UV radiation is identified as the root cause if the experimentally found m/z values now match the expected theoretical ones. Light-intensive diode array detectors are typically more critical here than variable wavelength detectors as they project the full bandwidth of their wavelength spectrum into the flow cell – light discrimination according to its energy takes always place by the polychromator behind the flow cell which acts as a kind of "photoreactor" in such a case. Replacing the diode array detector by a variable wavelength detector is typically an appropriate way to solve this.

2.3.4.3 In-Source Collision-Induced Dissociation

Gas phase fragmentation reactions are a fundamental element of mass spectrometric experiments. Imagine an analyte which is excited to vibrations in a well-controlled manner: this molecule will then selectively break into pieces at the weakest bonds, thus, creating a characteristic fragment spectrum. As long as this fragmentation reaction can be stimulated reproducibly, these fragmentation patterns provide you with a molecular fingerprint which tremendously helps to identify a chemical compound. And even if you cannot afford to run your entire LC/MS experiment in full-scan mode with parallel MS/MS fragmentation because you need highest sensitivity, the detection of as many known characteristic fragments as possible in SRM mode substantially increases the level of confidence for your compound confirmation. However, ESI and APCI are rather mild ionization principles transferring the analyte molecules into the gas phase as a whole, so fully intact and nonshattered – which is very different to electron ionization (EI, also earlier known as *electron impact ionization*) in GC/MS. The benefit is that you will be able to determine the molecular mass of the intact molecule, but you lose the chance to learn more about the structural properties and chemical nature of your analyte by fragmentation patterns. Sophisticated tandem-MS techniques, however, enable you to stimulate well-controlled decomposition conditions in the collision cell of a tandem mass spectrometer. The user typically knows about the most characteristic fragments of the target analyte, so not too many surprises are to be expected then. Many years of extensive research meanwhile allowed unveiling a huge set of decomposition reactions and their follow-ups in the gas phase [17, 18]; a very informative and comprehensive tutorial by Holcapek *et al.* is a valuable starting point for your own interpretation of small molecule fragmentations in API mass spectrometers [16].

In addition to these intended fragmentation reactions in a tandem MS, the user will always have the chance – or the risk – to shatter the analyte in an uncontrolled way, typically stimulated by unfavorable ion source or ion transfer conditions. As long as the analytes have not entered the final high vacuum section of the mass spectrometer, that is, while they still are in the ion source or the transfer section of the ion optics which come along with a staged pressure reduction, these ions will have to remain intact in an environment where their mean free path is only in the range of a few micrometers (approx. 50 µm at 1 mbar of ambient pressure) and not several dozens of inches (approx. 20 inches/500 mm for a vacuum of 10^{-4} mbar). A collision with excessive ambient gas molecules is very likely there, and the higher the collision impulse, the more you will see unwanted fragmentation taking place already in the entrance area of the ion source. You can master this process to some extent by a smart selection of acceleration voltages in the ion optics. High voltages, to be applied for example along the ion transfer capillary or to the skimmer electrodes, strongly accelerate the ions while they are travelling through the ion optics; this induces more effective collisions with residual gas molecules, thus, resulting in more fragment signals in the mass spectrum (also called *nozzle-skimmer dissociation*). In case of the MS method development, it is a useful approach to change the parameter settings for the ion source and the transfer section stepwise while monitoring the signal change and the spectrum quality accordingly to avoid excessive analyte fragmentation. However, you can also make use of this principle to artificially stimulate ion fragmentation and to learn more about unknown compounds even with rather simple and cost-effective instruments like single quad mass spectrometers.

A very comprehensive table on the generation of fragments out of various functional groups can be found in the literature [16]. We will briefly discuss the behavior of alcohols, aldehydes, and carboxylic acids as representative examples for frequently occurring fragmentation reactions. These compounds classes have a strong heteropolarity of the carbon–oxygen bond in common due to the high electronegativity of the oxygen atom. One immediate result of this is the neutral loss of water; the loss of carbon oxides is another one, depending on the chemical composition of the analyte. Once protonated in positive mode, alcohols preferably split off water ($R-OH_2^+$ from R–OH), thus, generating a R^+ fragment (equaling $[M + H - H_2O]^+$) which is lighter than the expected ion of the intact molecular ion by nominal 18 Da. Aldehydes lose carbon oxide (CO), ending up in a fragment ion $[M + H - CO]^+$. Aliphatic carboxylic acids typically lose the thermodynamically very stable carbon dioxide after protonation, while aromatic carboxylic acids under the same conditions preferably "only" split off water, leaving us with an acylium cation for detection.

2.3.4.4 Contaminants Eluting from the Instrumentation

Nevertheless, there will still be many situations where a good knowledge of gas phase fragmentation reactions and chemical expertise will not help to explain the existence of prominent mass signals in your MS spectrum: for instance, the sheer amount of mass signals being so high that they cannot be deduced by fragmentation reactions, or mass interferences not only showing up as one individual signal but as a series with distinctive patterns. The root cause here can be very trivial – contaminants are eluting from your chromatography or the mass spectrometer hardware. If your MS was thoroughly cleaned recently, then the (U)HPLC system and any fluidics connected with it would be blamed for that, and the potential contamination sources are numerous:

- *Bleeding* of a separation column being either old or unsuitable for MS detection results in an increased elution of the stationary phase bonding which is split off the carrier material surface and leads to an increased noise in the LC/MS chromatogram.
- *Plasticizers* are ubiquitous in pretty much all modern plastic materials, from sample vials over solvent lines, piston seals to filter frits, and many other pieces. Very common plasticizers are phthalates which can rapidly be identified based on their characteristic masses (m/z 279, 391, 413, 429, 454 and many more). Also lubricants and separating agents like erucamide (m/z 338, 360) can frequently be observed.
- *Perfluorinated compounds* can be leaching from perfluorinated polymers like the widely used polymer PTFE (Teflon™). PTFE is very popular for any fluidic device, and so for LC instruments as well, due to its high resistance against all kinds of chemical attacks. Typically, solvent lines and the plane or hollow-fiber membranes of the vacuum degassers are made of it. Many instrument manufacturers list the types of materials used for wetted parts in their device datasheets. However, it is mostly the LC/MS analysis for per- or polyfluorinated compounds (PFCs) or perfluorinated organic acids (PFOAs) which interferes with leaching fluorine-containing compounds, so most LC/MS applica-

tions are not adversely affected. Dedicated LC/MS instrumentation for PFC or PFOA analysis can be built from modified regular UHPLC systems; some vendors offer dedicated conversion kits for use in such cases. Fluorinated compounds reveal themselves indirectly in the mass spectrum; the naturally occurring fluorine (^{19}F) is isotopically pure – all other isotopes are of artificial origin. Hence, the isotope pattern of fluorinated compounds differs from the typical isotope intensity distribution of common organic compounds – the intensity of the monoisotopic mass signal in small organic molecules is typically overemphasized in the presence of fluorine.

- *Polyethers* like *polyethylene glycols* and *polypropylene glycols* (PEG and PPG, respectively) are another frequent contamination type which can be identified quickly due to their characteristic MS spectrum pattern. They are pretty ubiquitous in plastic materials, but they can also be introduced into the MS by working sloppily with disposable gloves or skin care products. These compounds never show up with only one single mass signal but always come with a very characteristic polymer distribution pattern [22]. The mass distance of the monomeric units are $\Delta m/z = 44$ for PEGs and 58 for PEGs which immediately reveals the chemical nature of these contaminants.

- *Polysiloxanes (silicones)* are core ingredients of many modern high-performance oils and vacuum grease. Oil vapor traces from the rough vacuum pump leaking into the mass spectrometer produce characteristic signals of, for example, m/z 371, 445, or 519. In such a case, the ion optics, like focusing multipoles or ion funnels, should be checked for cleanliness. A continuous stream of oil mist entering the MS can lead to a razor-thin oil film coating the metal surfaces of the ion guides or, in some cases, on the mass analyzer over weeks and months. This results not only in contaminant signals in your mass spectrum but also in a measurable sensitivity loss which makes a thorough and extensive cleaning mandatory. But how would it happen anyway that oil mist from the rough pump(s) could enter the MS interior in a way that also comes along with ion generation? Well, the most obvious reason for this is a nonideal installation of the various exhaust hoses of your MS. For convenience, the exhaust tubes of the vacuum pump(s) and of the ion source drainage are frequently tied together into the same lab exhaust ventilation nozzle. But by doing so, you allow the vacuum pump exhaust to diffuse backward into the MS ion source and further down into the mass spectrometer. Installing the pump exhaust and the ion source draining tube into different connectors of the lab ventilation with a distance of 1.5–3 ft (0.5–1 m) in-between is a very simple and effective solution to that issue.

- *Metal ions* can be a great origin for the generation of larger gas phase adducts with your target analyte molecules; alternatively, metal ions can also react with parts of your sample and, thus, inhibit the detection of compounds. We already discussed the formation of alkali metal adducts in Section 2.3.4.1. A stainless steel LC fluidics or massive hardware defects in a biocompatible UHPLC system, for example, a damaged injection valve, can result in a propagated release of iron ions and, thus, in iron/analyte clusters in the mass spectrum; these clus-

ters reveal themselves very quickly due to their multicharge state and the isotope pattern of iron which significantly deviates from those of the usual elements in organic matter like carbon, hydrogen, nitrogen, and oxygen. Biochemical applications are particularly prone to issues created by heavy metals in the mobile phase. A lot of biological compounds tend to form either precipitates or nonvolatile aggregates with iron, or alternatively to irreversibly adsorb on iron surfaces. A frequently described phenomenon is the "vanishing", so the nondetectability of phosphorylated peptides and proteins in a separation system with a stainless steel fluidics. To avoid this, bioanalytical applications are preferably run on instrumentation with an iron-free fluidics which can be made of titanium, biocompatible metal alloys like MP35N, or PEEK, with the latter being not very pressure-resistant (tubing typically up to 5000–6000 psi/350–400 bar) and, thus, not being suitable for UHPLC applications.
- *Dissolved residual gases* in the mobile phase, however, typically result in an unstable spray and/or interfering spikes in the LC/MS chromatogram and the mass spectra rather than modifying the analyte mass signals.

The aspects discussed here can only represent a selection of some very popular phenomena, but this list is far from being a comprehensive compilation of known interference signals [19] Already some years ago, Keller *et al.* published an excellent and highly detailed, tabulated collection of all literature-known MS contaminants that were known at the time [20]. Most MS vendors discuss this topic very openly as well and compiled various collaterals on MS contamination sources [21, 22], and last but not least the internet offers various public data search engines on the matter. One example to be mentioned is the *MaConDa* (*Ma*ss spectrometry *Con*taminant *Da*tabase) database maintained by the University of Birmingham [23] which allows one to search for over 200 contaminants based on accurate mass, compound class, and mass spectrometer type (operation principle and manufacturer), featuring also many literature references.

2.3.5
Instrumental Reasons for the Misinterpretation of Mass Spectra

Finally, we will investigate some selected instrumental reasons that can lead to the misinterpretation of mass spectra. As discussed earlier, all MS types have their strengths and weaknesses that also affect the quality of the analytical result. Let us discuss these in the following on three scenarios.

2.3.5.1 False Mass Assignment Depending on Ionization Principle
As we already saw in Section 2.3.4.1, the selected ion creation principle determines which m/z value is shown in your mass spectrum for an unknown analyte species. The most frequently used ionization process by charge transfer, that is, proton association or distraction (used in ESI and APCI), does not result in the molecular mass of the intact molecule being measured, but in a m/z value differing by one proton mass (or multiple proton masses at a respectively corrected fraction

of the intact molecule mass for multicharged molecular ions). Other processes like APPI or EI (hardly used in LC/MS) which can also create ions by transferring electrons instead of protons lead to a measured m/z value which deviates from the theoretical value of the intact molecule only by the much smaller mass amount of an electron. With ESI, APCI and APPI, the formation of adducts with alkali metals and/or residual solvent molecules can lead to misinterpreting a numerical m/z value as representing a $[M + H]^+$ ion species which in fact would be, for instance, a $[M + H_2O + Na]^+$ ion instead. A thorough look at your mass spectrum can be very helpful here, as many adduct species coexist with others, with characteristic m/z differences between the various ion aggregates. So if you find a new m/z value which you may take for a $[M + H]^+$ ion, just check for further signals with m/z differences to the first one of, say, $["M + H" + 22]^+$, $["M + H" + 38]^+$, or $["M + H" + 41]^+$ – with these being a sodium, potassium, and ACN/proton adduct, respectively, this series can additionally confirm (or disprove) your originally assumed mass assignment.

2.3.5.2 False Interpretation Due to Poor Mass Resolving Power

Another reason leading to a mass signal misinterpretation is a poor or at least inappropriate mass resolution. Imagine two different analyte species having only very small differences in their m/z values and arriving in the mass analyzer simultaneously; a low-resolving mass spectrometer will not be able to sufficiently discriminate the two different masses. The resulting mass spectrum will, thus, show only the envelope curve for the two different mass patterns, and the peak maxima of this envelope function do not necessarily have to be identical to the mass signal maxima of the mass spectrum of each individual compound (see Figure 2.9). Also, low-resolution mass spectrometers will not be able to resolve higher charge states; we know that the m/z distance between the isotope pattern signals of a compound equals the $1/n$th fraction of the charge state n, which means that for analyte ions with three or more charges the isotope pattern cannot be resolved appropriately with low-resolution mass spectrometers (Figure 2.10). In summary, co-eluting contaminants or impurities could not be identified as such, or mass signals are erroneously assigned to the wrong compounds. Thus, what minimum resolving power of your mass spectrometer must have depends – in addition to your budget of course – on the sample complexity, the relative m/z differences to be discriminated, and the quality of your separation. The better your chromatography, the more unambiguous the interpretation of your mass spectrum will be in the end. As a general recommendation, a mass spectrometer for the determination of accurate masses should have a resolving power R of 10 000–15 000 at least; the scientific literature typically defines R of 10 000 as the minimum for high-resolution and R of 100 000 for ultra-high resolution mass spectrometry [6].

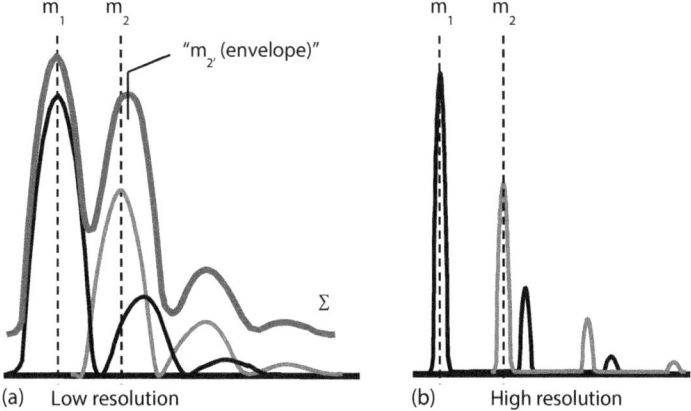

Figure 2.9 Mass spectrum of two co-eluting compounds m_1 and m_2, once measured with a low-resolving (a) and a high-resolving (b) mass spectrometer.

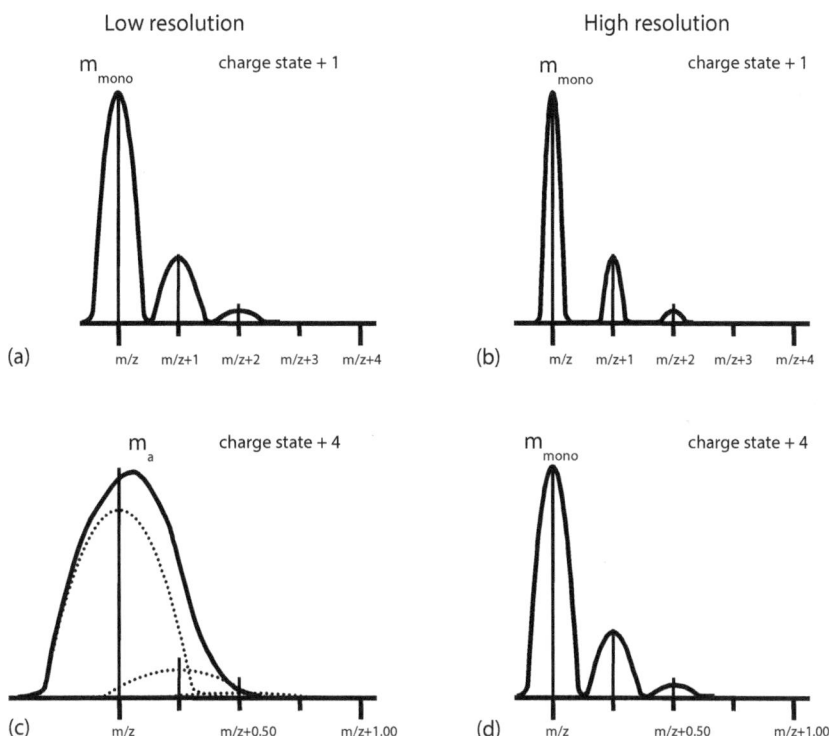

Figure 2.10 Mass spectrum of a single (a, b) and fourfold charged (c, d) compound, once measured with a low-resolving (a, c) and a high-resolving (b, d) mass spectrometer.

2.3.5.3 False Mass Determination Due to Inappropriate or Unstable Experimental Conditions

It is obvious that a mass determination is always achieved by comparing your instrumental MS data for the analyte with the mass signals of known calibration standards. Thus, the quality and the long-term stability of the mass calibration are critical for a reliable mass measurement. Expired or contaminated reference standards with partially degraded ingredients should of course not be used for calibration anymore. An undefined number of mass signals during the mass calibration process compromises the correct mass assignment and impairs or even blocks software tools like autotune algorithms. Once successfully calibrated, the quality of the mass axis calibration needs to be verified regularly. As discussed earlier, TOF instruments are particularly prone to drifts in the mass axis calibration even on short-term periods of (much) less than 1 h. Hence, an internal mass calibration by a continuous calibrant infusion is essential for a reliable mass determination. Mass spectrometers of the ion trap design (QIT, LIT, FTICR, Orbitrap) are additionally influenced by the spatial density of the ions stored in the electromagnetic field cage of the ion trap. The circulating ion packages in the trap act as so-called *space charges* which induce additional electrical fields, shielding, and, thus, locally distorting the external electromagnetic trapping field. Higher charge densities, which means high amount of analyte ions in the trap cell and/or multiply charged ion species, translate then into a pronounced shift of the resonance conditions for your circulating ion packages, which significantly impairs mass accuracy and mass resolution. An overfilled ion trap will then give you shifted and thus falsified mass signals. In contrast, the fewer ions you trap, the more you will lose sensitivity. The ideal filling degree for an ion trap is dynamically calculated in real time by modern MS control software – feature names here are *ICC, AGC* or others, depending on the MS vendor. Very concentrated samples, however, cannot always fully be intercepted by the control algorithms and, thus, will still lead to a short-time overfilling of the ion trap. Due to their design principle with a stretched longitudinal cell construction, linear ion traps (LIT) typically suffer less from space charge effects than circular traps (QIT).

2.4 Conclusion

For almost two decades, the coupling of liquid chromatography (LC) and mass spectrometry (MS) has been successfully commercialized. With the first instrument generation being true divas requiring in-depth expert knowledge, nowadays this technology has reached a fairly mature development state which substantially lowered the entry barrier to this technique. This results in many robust LC/MS solutions being established on the market, and LC/MS is penetrating more and more the field of routine applications, as users do not have to adopt a high amount of expertise to create quick and reliable results. Within these 20 years of growth and evolution, not only new mass spectrometry technologies like the Orbitrap have seen the light of day, but liquid chromatography has also made a big step ahead by moving from HPLC to UHPLC with much higher separation efficiencies and shorter run times. So from a bird's perspective, LC/MS has evolved to a very powerful analytical tool which is surprisingly easy to use given the high complexity of the technologies involved. But nevertheless, mass spectrometry is not the analytical all-purpose weapon as it is advertised in some cases, and it probably will never be. Key to the highest analytical benefit of an LC/MS installation is the thorough mutual physicochemical optimization of the LC and the MS world; this chapter of the book will hopefully contribute to a better understanding of the technical needs and to avoid the most common pitfalls. As scientific progress continues to move forward, several new technological territories will certainly be entered in the future, be it for even higher speed and resolution, for instrument miniaturization, or for enhanced usability by new and powerful software tools. However, one thing is for certain: It is not only the liquid phase separation that benefits from the mighty analytical information created by mass spectrometry. Also MS technologies will massively fall short on their potential without a thoroughly optimized chromatography upfront. For the foreseeable future, both concepts, chromatography and mass spectrometry, will continue to depend on each other.

2.5
Abbreviations

AP(C)I	atmospheric pressure (chemical) ionization
CAD	charged aerosol detection
CID	collision-induced dissociation
DAD	diode array detector
ECD	electrochemical detection
EI	electron ionization (also: electron impact ionization [obs.])
EIC	extracted ion chromatogram
ELSD	evaporative light scattering detector
FA	formic acid
FIA	flow injection analysis
FT	Fourier transformation
GDV	gradient delay volume
HILIC	hydrophilic interaction liquid chromatography
HPG	high-pressure gradient pump
HR/AM	high resolution/accurate mass
HTS	high throughput screening
HV	high voltage
ICR	ion cyclotron resonance
I.D.	inner diameter
LIT	linear ion trap
LPG	low-pressure gradient pump

MRM	multiple reaction monitoring
MSD	mass selective detector
PEEK	poly(ether ether ketone)
PEG	Poly(ethylene glycol)
PFC	perfluorinated compounds
PFOA	perfluorinated organic acids
PPG	poly(propylene glycol)
QC	quality control
QIT	quadrupole ion trap
RP	reversed phase
SIM	single ion monitoring
SST	stainless steel
SRM	selected reaction monitoring
TFA	trifluoroacetic acid
TOF	time of flight

References

1 Rogatsky, E., Zheng, Z., and Stein, D. (2010) *J. Sep. Sci.*, **33**, 1513–1517.
2 Thermo Fisher Scientific: "Viper™ Fingertight Fitting, Available: https://www.thermofisher.com/de/de/home/industrial/chromatography/liquid-chromatography-lc/viper-fittings.htm, (accessed 10 March 2017).
3 Agilent Technologies (2015) "A-Line Fittings", Available: http://www.agilent.com/en-us/products/liquid-chromatography/lc-supplies/capillaries-fittings/a-line-fittings, (accessed 10 March 2017).
4 MicroSolv Technology Corporation: Sure-Fit Connectors, Available: https://www.idex-hs.com/fluidic-connections/fittings/coned-fittings/surefittmconnectors.html, (accessed 10 March 2017).
5 IDEX Health & Science (2016) MarvelX, Link: https://www.idex-hs.com/marvelx, (accessed 10 March 2017).
6 Holcapek, M., Jirasko, R., and Lisa, M. (2012) *J. Chromatogr. A*, **1259**, 3–15.
7 de Hoffmann, V. and Stroobant, V. (2009) *Mass Spectrometry – Principles and Applications*, 3rd edn, John Wiley & Sons Ltd, Chichester, West Sussex.
8 Rodriguez-Aller, M., Gurny, R., Veuthey, J.-L., and Guillarme, D. (2012) *J. Chromatogr. A*, **1292**, 2–18.
9 Schappler, J., Nicoli, R., Nguyen, D., Rudaz, S., Veuthey, J.-L., and Guillarme, D. (2009) *Talanta*, **78**, 377–387.
10 Hopfgartner, G., Bean, K., and Henion, J. (1993) *J. Chromatogr.*, **647**, 51–61.
11 Asperger, A., Efer, J., Koal, T., and Engewald, W. (2001) *J. Chromatogr. A*, **937**, 65–72.
12 Thermo Fisher Scientific: LC/MS – Solvent selection; slide 8, Link: https://www.thermofisher.com/content/dam/tfs/Country%20Specific%20Assets/ja-ja/CMD/GCMS/faq/docs/technique/LCMS-basic-Choice-of-the-solvent-JA.pdf, (accessed 26 June 2016).
13 Dams, R., Benijts, T., Günther, W., Lambert, W., and De Leenheer, A. (2002) *Rapid Commun. Mass Spectrom.*, **16**, 1072–77.
14 Hoffmann, T. and Martin, M.M. (2010) *Electrophoresis*, **31** (7), 1248–1255.
15 Gosetti, F., Mazzucco, E., Zampieri, D., and Gennaro, M.C. (2010) *J. Chromatogr. A*, **1217**, 3929–3937.
16 Holcapek, M., Jirasko, R., and Lisa, M. (2010) *J. Chromatogr. A*, **1217**, 3908–3921.

17 McLafferty, F.W. and Turecek, F. (1993) *Interpretation of Mass Spectra*, University Science Books, Mill Valley.
18 Smith, R.M. (2005) *Understanding Mass Spectra: A Basic Approach*, John Wiley & Sons, Inc., Hoboken.
19 Guo, X.H., Bruins, A.P., and Covey, T.R. (2006) *Rapid Commun. Mass Spectrom.*, **20** (20), 3145–3150.
20 Keller, B.O., Suj, J., Young, A.B., and Whittal, R.M. (2008) *Anal. Chimm. Acta*, **627** (1), 71–81.
21 Agilent Technologies: What are the common contaminants in my GCMS. Link: http://www.agilent.com/cs/library/Support/Documents/FAQ232%20F05001.pdf, (accessed 10 March 2017).
22 Waters Corporation: ESI$^+$ Common Background Ions. Link: http://www.waters.com/webassets/cms/support/docs/bkgrnd_ion_mstr_list.pdf, (accessed 26 June 2016).
23 MaConDa: Mass spectrometry Contaminant Database, Link: http://www.maconda.bham.ac.uk/search.php, (accessed 10 March 2017)

3
Aspects of the Development of Methods in LC/MS Coupling

T. Teutenberg, T. Hetzel, C. Portner, S. Wiese, C. vom Eyser, and J. Tuerk

3.1
Introduction

In this chapter, we would like to explain and comment on a few of the most important observations with respect to the development of methods for LC/MS coupling. We will focus on the aspects of the interface between LC and MS which have been described in other textbooks, but have not been presented in the specific context of LC/MS coupling. At points where the more interested reader might profit from deeper deliberation, they are directed to the relevant special literature.

We assume that readers of this chapter are already versed in the basic terms used in liquid chromatography and mass spectrometry. This chapter is based on a DIN norm from the field of environmental analysis, which provides guidelines for the "determination of selected pharmaceutical ingredients and other organic compounds in water and waste water" [1]. DIN 38407-47:2015-07 is ideally suited as a general guide for discussion because although there are concrete recommendations for LC/MS analysis, it is well known that there are often many different ways to solve a problem. A conversion of the German norm DIN 38407-47 to the international standard ISO 21676 is under preparation at the moment [2]. Moreover, the less experienced user will probably be uncertain as to the amount of work involved in achieving the goals.

Even though this chapter has been written from the point of view of an environmental analyst, much of the content is transferable to many areas of the life sciences. After a short overview and categorization of the different analysis strategies (target analysis versus non-target analysis), we would like to discuss the question of a sensitive and pragmatic approach for method development in the context of LC/MS. This applies above all to the question of which parameters in chromatography and mass spectrometry should be optimized within the framework of the method developement and how the user can arrive at a satisfactory solution within a reasonable time span. The goal of the second part of this chapter is the discussion around data acquisition rates. Using concrete numerical examples, we are going to address the pitfalls that occur when highly efficient chromatographic procedures are to be coupled with diverse acquisition modes of the mass

The HPLC-MS Handbook for Practitioners, First Edition. Edited by S. Kromidas.
© 2017 WILEY-VCH Verlag GmbH & Co. KGaA. Published 2017 by WILEY-VCH Verlag GmbH & Co. KGaA.

spectrometer while a large number of compounds is also to be recorded during a single run.

3.2
From Target to Screening Analysis

3.2.1
Target Analysis

Nowadays, we generally differentiate between target and screening analysis. In target analysis, there is a fixed list of substances to be detected in the sample and their concentration determined. Reference standards are initially used to optimize the specific mass spectrometric and chromatographic parameters.

3.2.2
Suspected-Target Screening

In suspected-target screening, a search for expected substances is carried out. On first view, this definition seems to be inconsistent because, on the one hand, it is about a screening concept and, on the other, the detection of known or suspected substances. In contrast to target analysis, which primarily focusses on the quantification of the substances to be determined in a sample, in suspected-target screening a comparison to the reference standard is drawn to create a list of identification criteria such as the retention time, the precursor ion, the isotopic pattern, or the molecular formula that can be calculated from it, as well as the fragmentation pattern (MS/MS spectra). This information can be transferred into a database set up by the user just as well as in a free or commercial databases. Afterwards, the samples are measured and compared against a reference database. The higher the number of identification criteria, the higher the probability that the identity of the compound can be confirmed. Finally, a comparison with a reference standard solution can be performed to confirm the identified substance. This step makes sense if only a few identification criteria are fulfilled and the result does not seem plausible. At this point, we would like to direct the interested reader to further, more specialized literature which describes the workflow for suspected target screenings in detail [3].

3.2.3
Non-target Screening

In principle, in non-target screening there is no *a priori* information about the substances in the sample. The first indication for the identification of a substance is the determination of the accurate mass of the molecular ion. The elemental composition and a possible sum formula can be determined using the isotopic pattern. The next step, the allocation of a sum formula to a structural formula is considered extremely challenging. Therefore, it is essential that fragmentation spectra of the detected compounds are also measured in addition to the determination of the exact mass. For this purpose different high-resolution hybrid mass spectrometers such as quadrupole time-of-flight mass spectrometers or Orbitraps have been established. These offer the opportunity of fragmentation as well as the mere high-resolution MS full scan functionality for determining the exact mass of the quasi molecular ion. The precursor ion is selected, fragmented, and the resulting product ions are registered. The product ion spectra help to determine the structure of the substance. In addition, the use of meta-information such as the name and sum formula of the substance being searched for – or rather in the identification of transformation products or metabolites of the specific original substance – solubility, REACH data[1], bibliographical references, etc. are essential for making it possible to clearly allocate the accurate mass to a sum formula and finally even for making a structural suggestion. It is important to emphasize that non-target screening is extremely complicated and with regard to a reliable identification of an unknown compound very susceptible to errors. Here, we direct the interested reader to further literature which describes the workflow of non-target screening in more detail [4–7].

3.2.4
Comparable Overview of the Different Acquisition Modes

Table 3.1 contains a comparable overview of the workflows outlined in the previous sections. We would like to point out that the databases listed in the second column are not exclusively MS/MS spectra databases. The use of ChemSpider, for example, facilitates the pooling of information that are important for the characterization of an unknown compound. Furthermore, the databases listed here are rather unknown, so we would like to invite the reader to spend time gaining an overview of the information contained in these databases.

3.3
The Optimization of Parameters in Chromatography and Mass Spectrometry

3.3.1
Requirements and Recommendations for HPLC/MS Analysis Taking DIN 38407-47 as an Example

In the following sections, we are going to present the optimization of important separation and detection parameters for both chromatography and mass spectrometry. The discussion closely follows a current DIN norm draft for the mea-

---

1) REACH (Registration, Evaluation, Authorisation and Restriction of Chemicals) is a regulation of the European Union, adopted to improve the protection of human health and the environment from the risks that can be posed by chemicals, while enhancing the competitiveness of the EU chemicals industry. It also promotes alternative methods for the hazard assessment of substances in order to reduce the number of tests on animals.

Table 3.1 Overview of the different LC/MS workflows.

Target analysis	Suspected-target screening	Non-target screening
• Known substances • Direct comparison with reference substance • Identification using retention time, at least two specific mass transitions, precursor ion and/or fragmentation pattern • Quantification	• Suspected substances • Known sum formula and/or known fragmentation pattern • Verification by comparison of the precursor ion, sum formula and/or MS/MS spectra with self created, commercially or freely available databases such as CheLIST, ChemSpider™, DAIOS, Drugbank, HMDB, mzCloud™, Norman Massbank, Metfusion, Metlin, PPDB, Stoff-Ident, TOXNET [8–19]	• Not expected substances • No *a priori* information about the elemental composition • Determination of the sum formula • Narrowing down the possible chemical structures by the evaluation of the fragmentation pattern • Additional verification using *in silico* fragmentation[a] and/or transformation/metabolism by special software [20] • Statistical evaluation by principle component analysis (PCA)

a) The experimentally determined sum formula and product ion spectra are compared to simulated fragmentation pattern for further characterization of unknown substances.

surement of pharmaceuticals from water samples using LC/MS/MS. The individual aspects outlined in the norm are listed below and will be commented on later in the chapter by means of real world examples. We hope that users of LC/MS/MS processes will find some helpful reference points for a targeted optimization. In accordance with norm procedures in Germany for the determination of selected pharmaceuticals and other organic compounds in water and waste water, the following points should be considered:

- For chromatographic separation, use an appropriate HPLC column and optimize the separation of the analytes using gradient elution.
- Choose chromatographic conditions that will achieve optimum sensitivity for the mass spectrometric detection.
- Complete separation is not necessary as long as it is guaranteed that in the case of peak overlapping there is no interference in the quantitative determination.
- The shortest retention time should be at least three times the column void time.
- Substances that cannot be completely separated by mass spectrometry should be separated chromatographically. In such cases, the chromatographic resolution R should be at least 1.2.
- Properly adjust the injection volume so that there is no disruptive peak broadening or any interferences in the quantitative determination.
- Throughout six consecutive chromatograms, the standard deviation in the retention time should be no more than 0.03 min.

Alongside these recommendations, there are further remarks that should be considered:

- With regard to the sensitivity, it is advantageous to use a gradient with acetonitrile/water/acetic acid.
- For a constant linear velocity of the eluent, columns with a smaller inner diameter yield better sensitivity than those with a larger one.
- With a higher injection volume, for example, 1 mL, it is preferable to use column switching technology with the appropriate preconcentration columns.

Furthermore, the cited norm draws attention to special requirements in mass spectrometry:

- Following the defined conditions for chromatography, the optimum parameters for ionization should be chosen for every substance in the positive or negative mode taking the chemical characteristics of the substance into consideration.
- Choose substance-specific settings which, where possible, will result in two product ions for each analyte.

All necessary or relevant parameters that need to be considered for the development of an LC/MS method are named. In many textbooks on method development for chromatography, the general opinion is that the primary goal is the total separation of all compounds contained in the sample. This requirement comes from the fact that in many fields of the pharmaceutical industry, the UV detector is considered the "gold standard". The identification of co-eluting compounds is much more difficult using a UV detector than an MS detector. In this respect, the norm reflects the idea that mass spectrometric detection methods can more easily capture co-elution and chromatographic separation is desirable but not absolutely necessary. This then gives rise to the question of whether chromatographic separation is necessary at all if a mass spectrometer is used as the detector.

3.3.2
The Definition of Critical Peak Pairs in the Context of HPLC/MS Coupling

The demands on analytical processes in the field of environmental analysis have increased considerably. This is largely due to the increase in use of mass spectrometry as opposed to UV detectors because they facilitate the detection of organic compounds in very low concentrations. In many quantification and screening methods nowadays, significantly more than 100 components are included in one method. The fact that it is quite a challenge to carry out a complete baseline separation of about 50 compounds in one chromatographic run is illustrated in the following example which dates back to the deliberations of Calvin C. Giddings in 1983 [21]. He asked himself the question how high is the probability that a mixture of m components can be completely separated on an HPLC column with a given peak capacity. Let us assume that our sample contains 50 substances and the separation is to be carried out on a column with a peak capacity of 100;

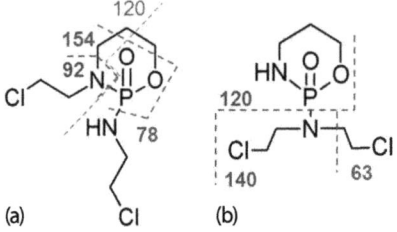

Figure 3.1 Structural formulae of (a) cyclophosphamide (CP) and (b) ifosfamide (IF).

therefore, 32 of the 50 compounds would co-elute. In fact, only 18 peaks would be baseline-separated. Therefore, the statement used in the norm – that a complete separation of all targeted analytes in the sample, or rather those specified in the method, is not absolutely necessary – can be regarded as a sensible criterion. The development of a chromatographic method in conjunction with a mass spectrometer should have the initial goal of only separating substances that are not distinguishable by mass spectrometry.

On the other hand, the question regularly asked is whether a chromatographic separation is necessary at all, as a perfectly clear distinction between co-eluting compounds is possible with a mass spectrometer. Using two examples, we would like to make clear that this is an unfortunate and regrettable misapprehension. The compounds presented in Figure 3.1 are structural isomers with an identical accurate mass.

In the case where a high-resolution mass spectrometer is used which is not able to generate product ion spectra, a mass spectrometric differentiation is not possible with the co-elution of these substances. In the case where a triple quadrupole mass spectrometer is used, a verification of the compounds despite co-elution might be possible on the basis of different mass transitions. The areas in the structural formulae marked with dashes and numbers show the characteristic "fragmentation positions" in the depicted molecules. As can be seen, in spite of the identical accurate mass there are many mass transitions to be selected. Ideally, the mass transitions chosen for identification and quantification should be different. If this is not the case because one or both of the mass transitions chosen for quantification and verification are identical for both compounds, it is mandatory to achieve chromatographic separation.

The substances shown in Figure 3.2 are epimers which only differ from each other in the position of the OH group. During co-elution of these compounds, a clear distinction can not be made using either a high-resolution mass spectrometer or a triple quadrupole mass spectrometer. A chromatographic separation in conjunction with the injection of the single reference standards for determining the retention time for each component is therefore crutial.

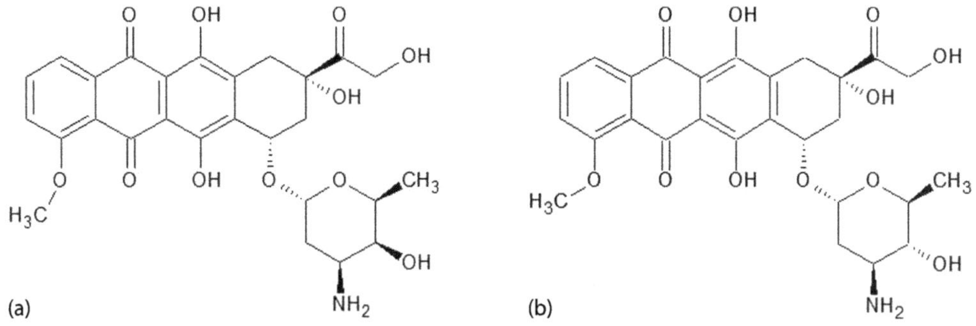

Figure 3.2 Structural formulae of (a) epirubicin and (b) doxorubicin.

3.3.3
The Separation of Polar Components from the Column Void Time

A further important criterion in the development of a multianalyte method for LC/MS is the separation of substances from the column void time. This presents a particular problem for extremely polar analytes that have only a very low or even no interaction with a classic reverse phase (RP) stationary phase, as at the column void time a significant signal suppression for all mass transitions is often observed. This phenomenon is visible in the chromatogram presented in Figure 3.3. This is a so-called matrix effect chromatogram, where the matrix, in this case an extract of house dust, is injected and after the column the analytes are passed into the

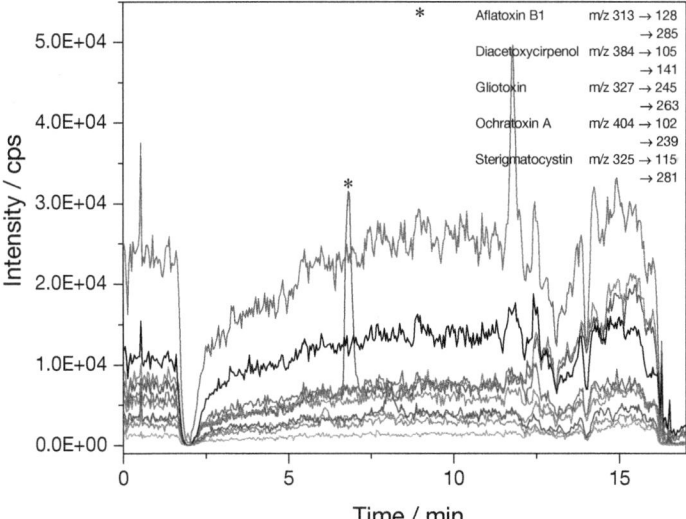

Figure 3.3 Matrix effect chromatogram of a house dust extract. The selected mass transitions of different mycotoxins (nonsmoothed raw data) are presented. For further explanations, see the text.

eluent flow via a T-piece. Recording of the target analytes takes place throughout the chromatographic run. Under optimum conditions, the intensity of every mass transition should be constant. However, exactly during the column void time the signals for all mass transitions almost completely break down, as at this time the salts found mainly in the matrix are eluting. Therefore, an important target criterion is to elute the polar compounds with a retention factor of > 2, or as it is stated in the norm, the shortest retention time should be at least three times the column void time.

As shown in the matrix effect chromatogram of Figure 3.3, particularly of the mass transition marked with an asterisk, a signal occurs at a retention time of around 7 min. This is an interference from the matrix which has an identical mass transition as the target compound, in this case aflatoxin B1. If the target compound eluted at the same retention time, this could lead to a false-positive result. In this case, a chromatographic separation would be essential. Hence, in order to increase the selectivity, it is advisable to measure at least two specific mass transitions per substance. When using a high-resolution mass spectrometer, at least one specific product ion should be measured.

3.3.4
Determining the HPLC Method Parameters Using the Example of the Separation of Selected Pharmaceuticals

The following section describes a case study for the development of a method containing 12 pharmaceuticals. It is outlined how sufficient chromatographic resolution can be obtained for critical peak pairs that require separation due to their mass spectrometric characteristics. Table 3.2 lists the names of the targeted

Table 3.2 A list of the target compounds with important physicochemical parameters [9].

Number	Compound	Molecular formula	Accurate mass (u)	pK_a ($T = 298.15$ K)	log D ($T = 298.15$ K, pH = 3)
(1)	5-Fluorouracil	$C_4H_3FN_2O_2$	130.0179	7.76/8.02	−0.65
(2)	Gemcitabine	$C_9H_{11}F_2N_3O_4$	263.0718	3.60/11.52	−3.34
(3)	Methotrexate	$C_{20}H_{22}N_8O_5$	454.1713	3.41/4.70	−2.95
(4)	Topotecan	$C_{23}H_{23}N_3O_5$	421.1638	8.00	−3.36
(5)	Irinotecan	$C_{33}H_{38}N_4O_6$	586.2791	11.71	−0.34
(6)	Ifosfamide	$C_7H_{15}Cl_2N_2O_2P$	260.0248	13.24	0.75
(7)	Cyclophosphamide	$C_7H_{15}Cl_2N_2O_2P$	260.0248	12.78	0.50
(8)	Doxorubicin	$C_{27}H_{29}NO_{11}$	543.1741	9.53	−2.86
(9)	Epirubicin	$C_{27}H_{29}NO_{11}$	543.1741	9.53	−2.86
(10)	Etoposide	$C_{29}H_{32}O_{13}$	588.1843	9.33	0.28
(11)	Paclitaxel	$C_{47}H_{51}NO_{14}$	853.3310	10.36	3.95
(12)	Docetaxel	$C_{43}H_{53}NO_{14}$	807.3466	10.96	2.46

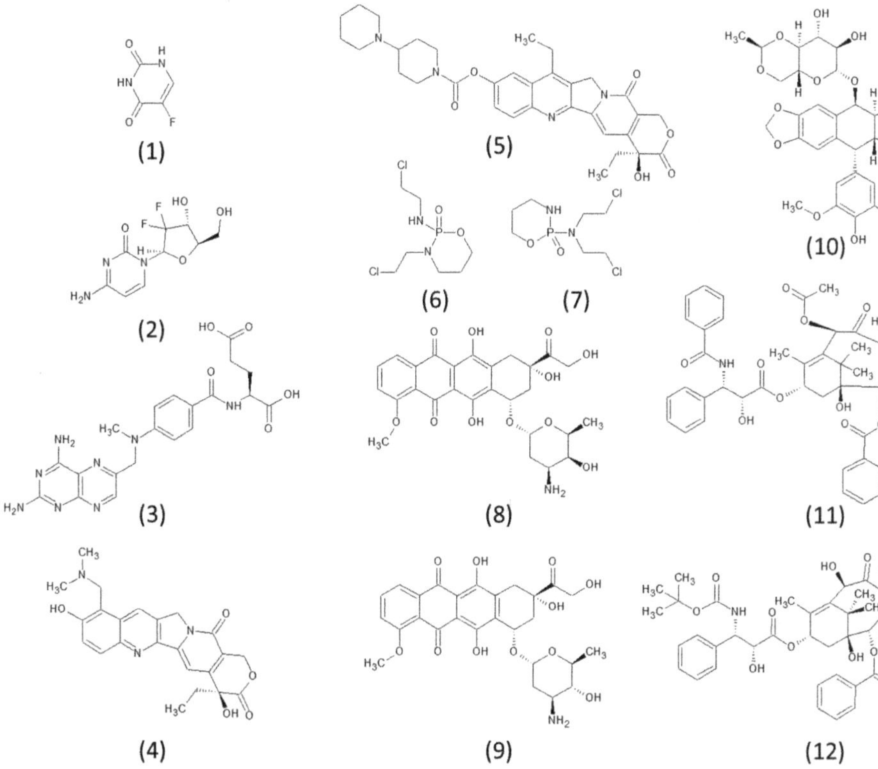

Figure 3.4 Structural formulae of the target analytes 5-fluorouracil (1), gemcitabine (2), methotrexate (3), topotecan (4), irinotecan (5), ifosfamide (6), cyclophosphamide (7), doxorubicin (8), epirubicin (9), etoposide (10), paclitaxel (11), and docetaxel (12).

compounds including important physicochemical data, while Figure 3.4 contains the structural formulae. The listed substances cover a broad spectrum of polarities which can be understood on the basis of the log D values. 5-Flourouracil and gemcitabine are polar enough to elute with or very close to the column void time on a classical RP stationary phase, while the other components exhibit a higher retention. Nevertheless, the log D values are not an indication of the actual elution sequence. Besides cyclophosphamide and ifosfamide, doxorubicin and epirubicin, the two taxanes, paclitaxel, and docetaxel should also be separated chromatographically. Although the two latter compounds exhibit different mass transitions and accurate masses, ion suppression was observed during co-elution.

The question that now arises is how to choose the "correct" or "appropriate" column. This seems to be an almost impossible task as there are probably around 1000 different stationary phases. Consequently, many users give up and simply use a column that is available in the lab. Such a pragmatic approach is certainly not wrong as the development of multianalyte methods always implies making compromises. However, it is often observed with such an approach that critical peak pairs are either insufficiently or not at all separated, as the selectivity of the stationary phase or the phase system is not adequate in most cases.

In order to limit the range of possibilities in terms of finding an appropriate stationary phase, the separation should be first optimized using RP chromatography. As stated in the norm, with regard to mass spectrometry the preferred eluents are those containing acetonitrile. During method development, methanol should also be used. Whereas acetonitrile is an aprotic solvent, methanol is considered protic and, therefore, closer to water as a solvent. This results in significant differences in the interactions between the organic solvent, the analytes, and the stationary phase, which in turn influence the selectivity. In the norm, it is recommended that acetic acid should be used as a mobile phase additive. However, in many cases formic acid might be more suitable. In general, when using additives, which are also used to enhance the ionization, it is important to ensure that no poorly soluble and nonvolatile substances are generated when the mobile phase is vaporized in the ion source. During positive ionization, one should also consider the possible generation of adducts, for example, $[M + NH_4]^+$ or $[M + Na]^+$, in addition to the quasi molecular ion $[M + H]^+$.

The next important parameter is the temperature of the stationary and mobile phases. The temperature exerts a significant influence on the separation by affecting the selectivity [22, 23]. Terefore, it makes sense to include at least two temperature levels in the optimization of the separation [24]. In fact, it is recommendable to carry out the optimization at 30 and 50 °C.

As the slope of the solvent gradient also influences the peak capacity and selectivity, all runs should be carried out with two different gradient slopes. As a general rule, the lower the slope of the gradient, the higher the resulting peak capacity.

The next step is to consider which stationary phases should be used for method development. Within the scope of a scientific study, which is the basis of the following conclusions, a comprehensive selection was assessed [25]. We are very aware that the number of stationary phases included in the study that are listed in Table 3.3 is very high. Nevertheless, it is advisable to have several columns available which exhibit fundamental differences in their selectivity. Besides the more classic
C-18 phases, this includes perfluorinated phases or those with a phenyl group. In addition, phases with an embedded polar group offer the possibility to carry out separations with a high water content. The number of phases included in a column screening depends not only on their availability in the laboratory but also on the amount of time avaiable for method development. Several manufacturers offer special kits for method development which contain four to six stationary phases of different selectivity. Independent of the number of columns the following screening parameters which can be seen in Figure 3.5 should be considered.

Formic acid (0.1%) is added to all eluents. A linear gradient from 5 to 95% B organic solvent (% B) was used. In some cases starting the gradient at 1% or even 0% can be useful, especially when extremely polar analytes are to be analyzed. It

Table 3.3 A list of the stationary phases selected for the screening.

Stationary phase	Modification	Particle	End capping	Particle diameter (μm)	Pore diameter (Å)
Agilent Zorbax SB	C-18	Fully porous	Sterically shielded	1.8	80
ChromaNik Sunshell RP-Aqua	C-28	Core–shell	Multistage	2.6	160
Macherey Nagel Nucleoshell RP 18plus	C-18	Core–shell	Multistage	2.7	90
Merck Chromolith FastGradient RP 18e	C-18	Monolith	End capped	Macro pores 1.5	Meso pores 130
Phenomenex Kinetex	C-18	Core–shell	Trimethylsilane	2.6	100
Phenomenex Synergi RP polar	Ether-linked phenyl phase	Fully porous	Polar end capping	2.5	100
Restek Raptor AR	C-18	Core–shell	Sterically shielded	2.7	90
Restek Raptor	Bi-phenyl		End capped		
Supelco Ascentis Express	C-18	Core–shell	Trimethylsilane	2.7	90
	C-8				
	CN				
	ES C-18				
	EPG Amide				
	Phenyl-Hexyl				
TCI Kaseisorb LC ODS-SAX Super	C-18 + Anion exchanger	Fully porous	End capped	3.0	120
Thermo HypersilGold	PFP	Fully porous	End capped	1.9	175
Waters Acquity BEH	C-18	Fully porous	End capped	1.7	130
Waters HSS T3	C-18			1.8	100
Waters XBridge	C-18			2.5	130
YMC Triart	C-18	Fully porous	Multistage	1.9	120

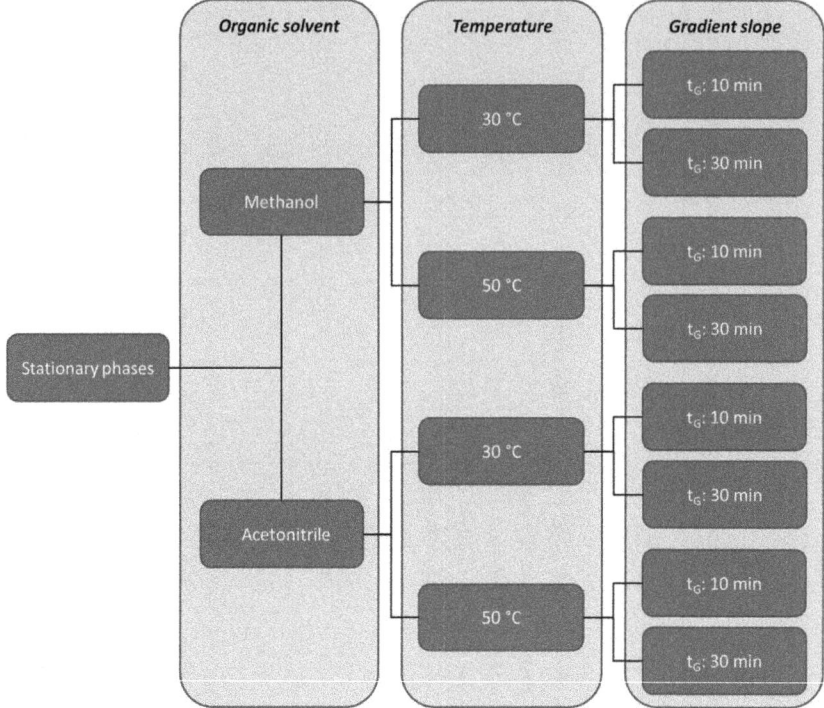

Figure 3.5 Graphic representation of the experimental design for determining the appropriate phase system.

should be noted that the water can be excluded from the pores on a hydrophobic reverse phase if the mobile phase does not contain any organic solvent. In the literature, this phenomenon is sometimes called dewetting or phase collapse and leads to a collapse of the hydrophobic alkyl chains on the surface. As a consequence, a complete loss of retention might be observed. In most cases, this effect can be reversed by rinsing the stationary phase with an organic solvent. Special phases that are identified by adding "aqua" or something similar to the column's name, make it possible to use only water as mobile phase. The end of the gradient can also be set to 99% B. However, the solubility of salts or polar additives decrease as the organic part increases in the mobile phase.

3.3.5
Carrying out Screening Experiments

3.3.5.1 Fully Automated Screening

The following section describes an approach that allows a largely automated screening – provided that a suitable HPLC system is available. An example of such a system is the Shimadzu's Method Scouting system which is based on the Nexera HPLC Series. This system enables the user to combine six HPLC columns with eight solvents. This results in 96 variations, generating a large amount of information in a very short time. The user only has to define the basic conditions such as the choice of columns and solvents, the setting of the oven temperature and the gradient slope. The time-consuming batch preparation is done by the software taking into consideration all relevant equilibration and flushing steps. This helps to reduce mistakes caused by insufficient knowledge of system or dwell volumes. Nevertheless, at least one run should be carried out as repeated injection in order to check the functionality of the system.

After the screening measurements, the chromatograms are evaluated automatically by the Agent Report Software. For this, a so-called evaluation factor is calculated for each phase system which is based on the number of detected signals multiplied by the resolution between adjacent peaks. The software extracts these parameters automatically for each of the phase systems being investigated. Via a link to a table calculation program, a result table is generated from the evaluation factors for the investigated stationary phases. However, the resulting evaluation factors can be distorted if very high-resolution values are obtained for certain peak pairs. For example, when many components in one chromatographic run co-elute and at the same time, only two substances are separated, with a relatively high resolution of, for example, $R = 5$, this can lead to a higher evaluation factor, although other phase systems would possibly be more appropriate. To avoid this problem, a maximum resolution can be defined in the software. This means that despite a real resolution of $R = 5$, the software calculates with the defined resolution of, for example, $R = 2$. This leads to a weighting of the results, and the risk of misinterpretation of the final results is reduced. All in all, this type of evaluation offers a simple possibility to identify the optimum phase system with minimal manual effort.

3.3.5.2 Manual or Partially Automated Screening

In case that a fully automated screening system is not available, the experiments have to be carried out individually in a series of measurements. An Agilent 1200 system was used in order to deal with all the parameters in the cited study. It was possible to place two columns in the column oven and to run the experiments semiautomatically. The selection of the columns followed the programmed acquisition method which includes a selection of the valve positions. Figure 3.6 shows the schematic layout.

The screening experiments are carried out following classic sequence processing practices. Thus, the different methods are performed one after the other. It is important to take all equilibration steps into consideration by programming a rinsing method before any solvent change. It is important to determine the system and dwell volumes exactly so that mistakes, which could distort the results, do not occur in the equilibration. At the end of the sequence, the next step is to connect two new stationary phases with different selectivities and the previously generated batch can be started again. Data evaluation is carried out manually.

Manual data acquisition requires much more time than the use of automated systems specifically set up for a screening. Purchasing of an HPLC system that

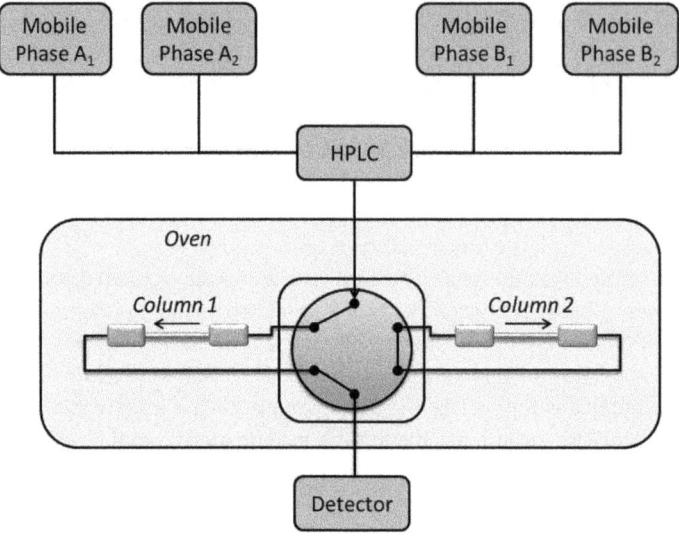

Figure 3.6 Schematic layout of the HPLC system used for partially automated screening.

will be used exclusively for method development is recommended when the lab often has to establish or adapt new methods. If only current methods or those methods described in the literature are being applied, it will not be advisable to invest in a separate screening system. The advantage of HPLC screening systems is their exclusive use for method development. If an optimized method is to be transferred to the system that is used for routine measurements, certain nasty surprises can arise. These problems are often due to differing system and dwell volumes. It has, therefore, proved advantageous to carry out fine tuning on the system used for routine measurements. We will be addressing this point again in Section 3.3.7.

3.3.6
Evaluation of the Data and Discussion of the Influencing Parameters

3.3.6.1 The Influence of the Stationary Phase

In an ideal case, multiple chromatograms, in which all compounds are separated with sufficient resolution, are already obtained during the screening runs. Using selected examples, we would like to give some tips on which separating systems are preferable under certain conditions, when the demand of a minimum resolution for the previously defined critical peak pair has been fulfilled. As this is a case involving UV chromatograms and not all substances could be detected well at a wavelength of 254 nm, a second UV chromatogram was integrated using a wavelength of 200 nm. On the basis of the available data, it would be possible to write a complete thesis about the different stationary phases integrated in the study. We are, therefore, going to limit the discussion on the influence of stationary phases to selected examples.

Taking a look at the columns listed in Table 3.3, it is visible that most of these phases are appropriate for the RP mode. The situation is probably similar in many labs: a silica based C-18 column of some specific manufacturer is used for most separation problems. Therefore, the first discussion needs to be on the influence that the length of the alkyl chain has on the quality of the separation.

In Figure 3.7, three chromatograms are presented that were obtained from silica-based reversed phases with acetonitrile as organic solvent. The separation presented in Figure 3.7a was carried out on a phase with a C-8 modification. A signal was obtained close to the column void time where 5-fluorouracil (1) and gemcitabine (2) are co-eluting. The vertical dotted line denotes the column void time multiplied by three. As there is a clear differentiation between the two substances in mass spectrometry both in the accurate masses and in their specific mass transitions, a chromatographic separation is not particularly necessary. However, both compounds elute very close to the void time so that co-eluting salts and polar compounds from the matrix could lead to a strong suppression of the signal (quenching or ion suppression) [26]. The remaining compounds elute with sufficient retention. The critical peak pair cyclophosphamide (6) and ifosfamide (7) can be separated with a resolution of 1.44. The components paclitaxel (11) and docetaxel (12) yield a similar resolution of 1.49. The chromatographic separation of this pair is only prophylactic as the differentiation between these compounds is possible both on the basis of the accurate masses and the characteristic mass transitions. Only the epimers doxorubicin (8) and epirubicin (9) are not sufficiently separated with a resolution of 1.10. All in all, the selectivity of the phase system is not sufficient to fulfill the minimum resolution of 1.2 for doxorubicin and andepirubicin as required by the cited norm for all critical pairs.

So let us take a closer look at the chromatogram obtained on the C-18 silica gel phase (Figure 3.7b). It is immediately clear that the first two eluting compounds can be separated, where 5-fluorouracil (1) elutes before gemcitabine (2). However, a sufficient separation from the column void time cannot be achieved in this case either. A satisfactory resolution for the isomers cyclophosphamide (6) and ifosfamide (7) was not obtained, whereas the resolution for doxorubicin (8) and epirubicin (9) is 1.37. The resolution for the taxanes that are last to elute is higher than 1.20 and therefore satisfactory.

The selectivity becomes even better when the separation is carried out on a C-28 modified RP stationary phase (Figure 3.7c). The first two eluting compounds are well separated ($R = 2.38$) but still elute very close to the column void time. The resolution of all three critical peak pairs is higher than or equal to the recommended value in the norm, where the components doxorubicin (8) and epirubicin (9) exhibit the lowest resolution ($R = 1.20$). The method development could now be stopped as an initial appropriate phase system that fulfills the minimum requirements regarding the chromatographic resolution has been identified. However, due to aging, the efficiency of the column can deteriorate more or less rapid-

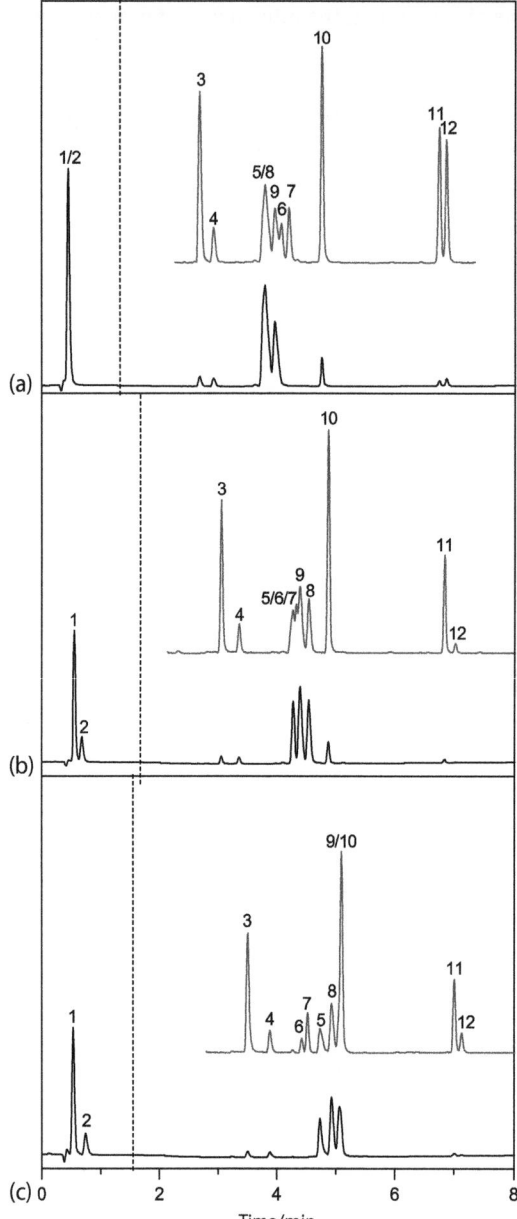

Figure 3.7 Comparative separation of 12 cytostatic drugs on (a) Supelco Ascentis Express C-8 (50 × 2.1 mm, 2.7 μm), (b) Restek Raptor ARC-18 (50 × 2.1 mm, 2.7 μm), (c) ChromaNik SunShell RP-Aqua C-28 (50 × 2.1 mm, 2.6 μm) columns; chromatographic parameters: temperature: 30 °C injection volume: 2 μL; mobile phase: A = water + 0.1% formic acid, B = acetonitrile + 0.1% formic acid; flow rate: 350 μL min^{-1}; detection: UV at 200 nm and 254 nm.

ly, so that the critical resolution can quickly fall short of 1.20, which means that a significantly higher resolution has to be the goal. In addition, the target criterion regarding a sufficient separation of the polar components from the void time was not achieved.

The next comparison, presented in Figure 3.8, comprises materials that are significantly different in their chemical structure. It can be seen that the first two eluting compounds 5-fluorouracil (1) and gemcitabine (2) cannot be separated on the RP-amide phase (Figure 3.8a). In contrast, the biphenyl phase (Figure 3.8b) and the mixed-mode phase (Figure 3.8c) are able to separate both components, where gemcitabine does elute as the first peak on the mixed-mode phase. In all three cases, there is no sufficient separation from the column void time. Consequently, for the polar compounds there is no significant difference in comparison to the classic alkyl-modified RP columns.

The resulting chromatogram obtained on the amide phase shows sufficient separation of all critical peak pairs with a minimum resolution for ifosfamide (6) and cyclophosphamide (7) of $R = 1.40$. Using the other two phase systems, however, no satisfactory chromatographic separation can be achieved. Only paclitaxel (11) and docetaxel (12) can be separated with a resolution of 1.20 if acetonitrile is used as the solvent. Therefore, the amide phase would also be suitable for separating the critical peak pairs with the required minimum resolution. Having said that, the separation of the polar compounds from the column void time is not sufficient either.

3.3.6.2 The Influence of the Organic Solvent

As already mentioned, screening experiments should also be carried out using methanol, as an improved spread of the elution bands in the gradient window has been observed. The reduced elution strength of methanol can, in certain circumstances, lead to a higher retention for the polar substances 5-fluorouracil and gemcitabine. To evaluate the influence of the organic solvent, Figure 3.9 shows separations carried out using the identical stationary phases as those in Figure 3.7 with all conditions constant, except that methanol is used instead of acetonitrile. On observing the polar substances, a clear increase in retention can be seen for gemcitabine (2) in comparison to the separation using acetonitrile (see Figure 3.7). Even on the C-8 phase a partly separation is now achieved between 5-fluorouracil (1) and gemcitabine (2). Nevertheless, the retention of the polar substances is still less than three times the column void time. All other components exhibit an increase in retention compared to the separation using acetonitrile, which can be explained by the reduced elution strength of methanol. The separation presented in Figure 3.9a, carried out on a C-8 phase, shows a minimum resolution of $R = 1.17$ for the epimers doxorubicin (8) and epirubicin (9), which means that the requirements of the norm are not completely fulfilled. Both the C-18 and C-28 phases are able to generate a separation of all critical peak pairs under the chosen conditions, the minimum resolution being $R = 1.23$. The significantly improved separation of the isomers cyclophosphamide and ifosfamide with a resolution > 2 is particularly of note. This example clearly proves that acetonitrile is not the only solvent to

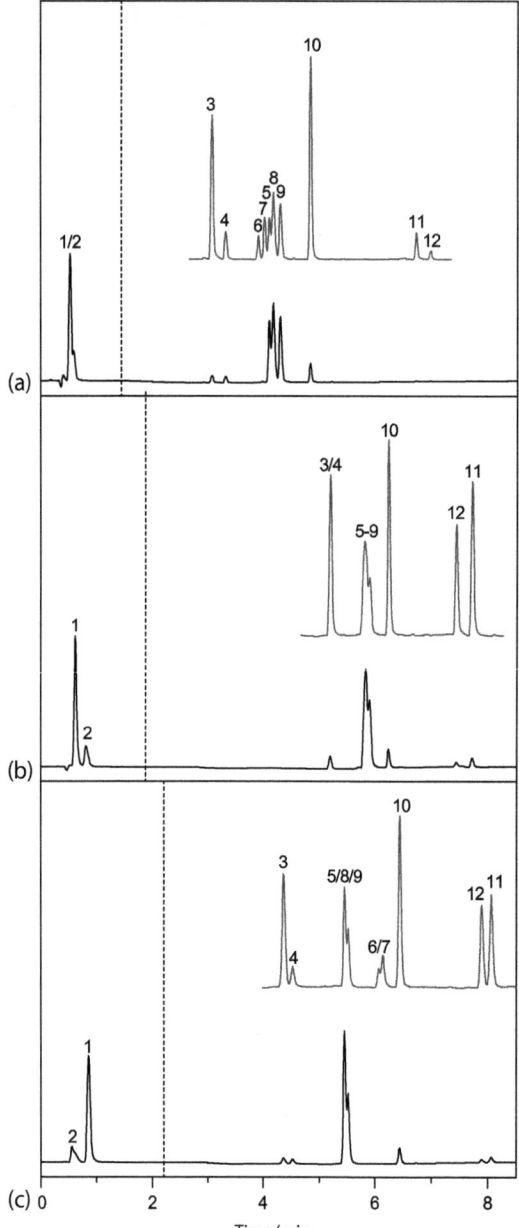

Figure 3.8 Comparative separation of 12 cytostatic drugs on (a) Supelco Ascentis Express RP-amide (50 × 2.1 mm, 2.7 μm), (b) Restek Raptor biphenyl (50 × 2.1 mm, 2.7 μm), (c) TCI Kaseisorb LC ODS-SAX Super (50 × 2.0 mm, 3.0 μm) columns; chromatographic parameters: temperature: 30 °C; injection volume: 2 μL; mobile phase: A = water + 0.1% formic acid, B = acetonitrile + 0.1% formic acid; flow rate: 350 μL min^{-1}; detection: UV at 200 nm and 254 nm.

3.3 The Optimization of Parameters in Chromatography and Mass Spectrometry | 91

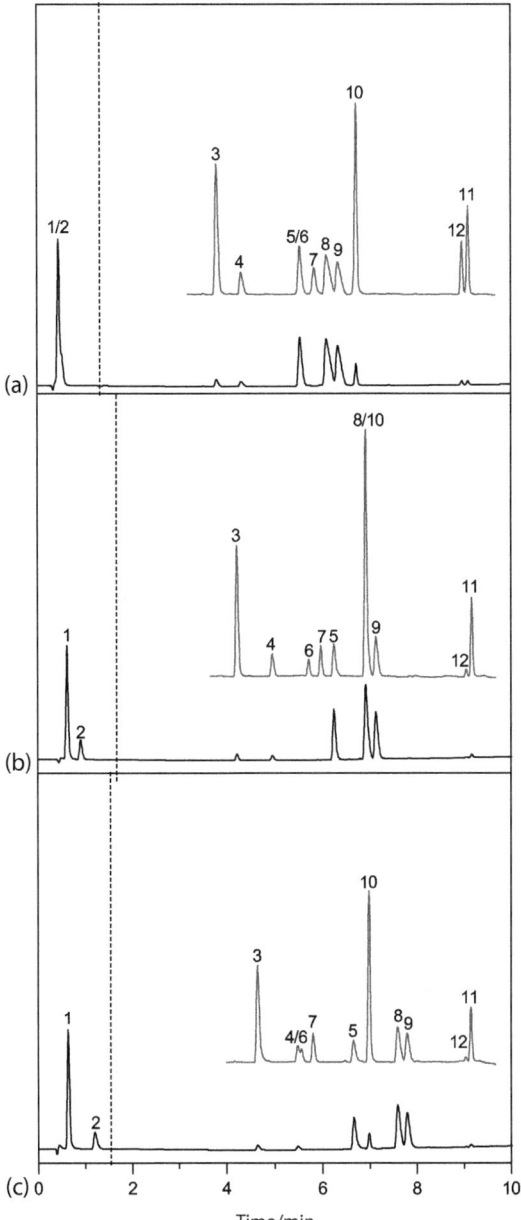

Figure 3.9 Comparative separation of 12 cytostatic drugs on (a) Supelco Ascentis Express C-8 (50 × 2.1 mm, 2.7 µm), (b) Restek Raptor ARC-18 (50 × 2.1 mm, 2.7 µm), (c) ChromaNik SunShell RP-Aqua C-28 (50 × 2.1 mm, 2.6 µm) columns; chromatographic parameters: temperature: 30 °C; injection volume: 2 µL; mobile phase: A = water + 0.1% formic acid, B = methanol + 0.1% formic acid; flow rate: 350 µL min^{-1}; detection: UV at 200 nm and 254 nm.

be used for method development. In general, methanol has the effect of expanding the elution window, thereby, improving the spread of the chromatographic bands. This is particularly advantageous when using older mass spectrometers that have a lower data acquisition rate, which will be discussed in more detail in Section 3.3.13.

Finally, in Figure 3.10, we have the comparative chromatograms for the alternative stationary phases from Figure 3.8, again using acetonitrile instead of methanol. On closer investigation, we see that in two cases for paclitaxel (11) and docetaxel (12) a separation yields no sufficient resolution. Only the biphenyl phase is able to generate a separation under the chosen conditions. It should be mentioned here once again that from the point of view of mass spectrometry, a separation of this substance pair is not absolutely necessary. Even though the biphenyl phase is advantageous for separating the taxanes, this phase is less suitable for separating the other critical peak pairs. In this regard, the amide and mixed-mode phases are superior to the biphenyl phase. A baseline separation is achieved for both the isomers and the epimers.

On the basis of these chromatograms, the recommendation is to perform the separation on the C-18 phase with methanol as the organic solvent, as this results in a significantly higher chromatographic resolution for the two critical peak pairs cyclophosphamide and ifosfamide, and doxorubicin and epirubicin than the minimum requirement stated in the norm for water quality analysis. Whether to use acetonitrile or methanol as the mobile phase depends not only on the influence on the quality of the chromatographic separation but also on the resulting pressure drop during the separation. In general, under identical chromatographic conditions during the solvent gradient for a binary system of water–methanol, a significantly higher maximum pressure is reached than for a binary system of water–acetonitrile [24]. This is of course the reason why higher chain alcohols such as ethanol or isopropanol are not widely used in RP chromatography as the viscosity is higher and the maximum pressure is, therefore, higher than for a binary system of water–methanol. Another alternative solvent for optimizing the selectivity is tetrahydrofuran (THF) [27]. This solvent plays a particularly important role for size-exclusion chromatography (SEC) of polymers, as they are not often soluble in methanol or acetonitrile. But using THF exhibits some clear disadvantages. Due to its good solvent properties, THF attacks plastics that can be found in the flow path of the mobile phase. In addition, it is difficult to obtain ultrapure THF, which is problematic for trace analysis using LC/MS. In addition, THF has a tendency to form peroxides and is generally mixed with stabilizing agents. On account of these disadvantages, we are not going to pursue any further reflections on the optimization of selectivity by using alternative solvents, here.

3.3.6.3 The Influence of Temperature

Temperature plays an important role with regard to selectivity of the separation. Moreover, it is apparent that the influence of the temperature on important physicochemical parameters, for example, viscosity of the mobile phase, the diffusion coefficient of the analytes in the mobile and stationary phases, the retention

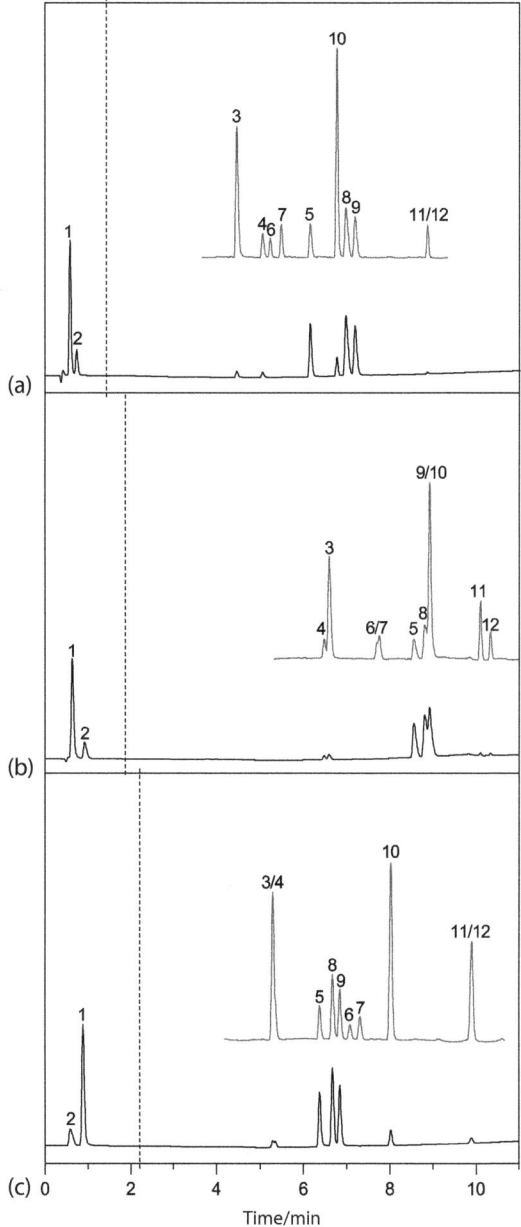

Figure 3.10 Comparative separation of 12 cytostatic drugs on (a) Supelco Ascentis Express RP-amide (50 × 2.1 mm, 2.7 μm), (b) Restek Raptor biphenyl (50 × 2.1 mm, 2.7 μm), (c) TCI Kaseisorb LC ODS-SAX Super (50 × 2.0 mm, 3.0 μm) columns; chromatographic parameters: temperature: 30 °C; injection volume: 2 μL; mobile phase: A = water + 0.1% formic acid, B = methanol + 0.1% formic acid; flow rate: 350 μL min^{-1}; detection: UV at 200 nm and 254 nm.

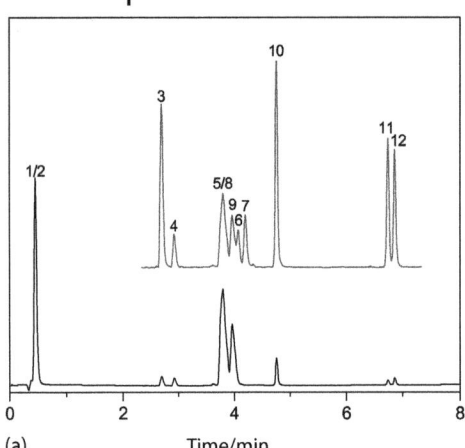
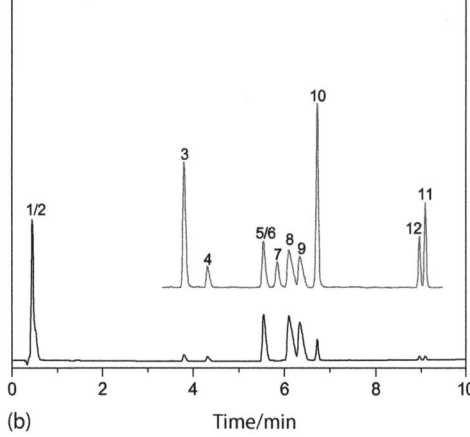
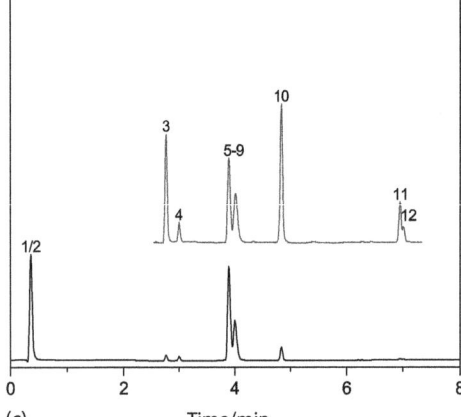
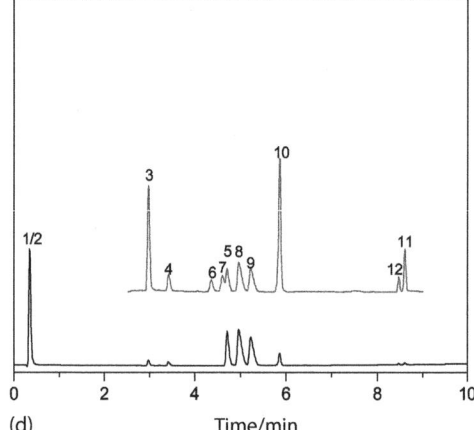

Figure 3.11 Comparative separation of 12 cytostatic drugs on an Supelco Ascentis Express C-8 (50 × 2.1 mm, 2.7 μm) column; chromatographic parameters: (a, b) temperature: 30 °C, 50 °C; injection volume: 2 μL; mobile phase: A = water + 0.1% formic acid, B = acetonitrile + 0.1% formic acid; flow rate: 350 μL min^{-1}; detection: UV a 200 nm and 254 nm; (c, d) temperature: 30 °C, 50 °C; injection volume: 2 μL; mobile phase: A = water + 0.1% formic acid, B = methanol + 0.1% formic acid; flow rate: 350 μL min^{-1}; detection: UV at 200 nm and 254 nm.

factor, is underestimated in many cases. In general, temperature can influence the selectivity positively or negatively. Figure 3.11 presents an exemplary case for the C-8 phase. Figure 3.11a,b shows the results of a separation using acetonitrile where the temperature was adjusted to 30 and 50 °C, respectively. In comparing the chromatograms, one can conclude that the signals that are partially separated at a temperature of 30 °C, completely co-elute at around 4 min when the temperature is increased to 50 °C. This is also true for the taxanes paclitaxel (11) and docetaxel (12) with a retention time of 7 min. In this case, the increase in temperature leads to a decreased resolution for all critical peak pairs. In contast, if methanol is used, the higher temperature leads to an improvement in the selectivity and resolution, as can be seen in Figures 3.11c,d, respectively. This is also reflected in the calculated resolution values. A minimal deterioration in the resolution can be seen for ifosfamide (6) and cyclophosphamide (7), whereas an improvement is achieved for doxorubicin (8) and epirubicin (9) from 1.17 at 30 °C to 1.53 at 50 °C. The resolution for paclitaxel (11) and docetaxel (12) is hardly influenced by temperature. Therefore, it is not possible to make a general statement about the influence of temperature on the corresponding resolution. This would always require specific consideration of the analyte in conjunction with the phase system being used.

3.3.6.4 The Influence of Gradient Slope

The chromatographic resolution can also be improved by changing the slope of the gradient because a low slope usually results in a higher peak capacity [28]. Looking at the two chromatograms being compared in Figure 3.12, it is clear that the longer gradient run time yields a better resolution.

In this case, it is also possible to obtain a baseline separation of all compounds with acetonitrile as the organic solvent. The disadvantage is the longer analysis time, which is the trade-off for obtaining a higher resolution. The decision regarding the method more suitable for routine applications is left to the user. As already mentioned, acetonitrile has some advantages over methanol. The acetonitrile–water separation presented in Figure 3.12a would be definitely a good alternative. After the elution of the last compound, the cycle time can be shortened by increasing the slope of the gradient to, for example, 95% within 1 min, in order to elute hydrophobic matrix components from the column. Once again a satisfactory separation of the polar compounds from the column void time could not be achieved. We will discuss this problem again in Section 3.3.10.1.

3.3.6.5 The Influence of the pH Value

The pH value is also an important parameter when trying to improve the selectivity as well as the robustness of a method. As far as LC/MS coupling is concerned, it is common practice to simply add ionization additives to the mobile phase in order to roughly adjust the pH value. Adding 0.1% acetic acid is enough to achieve a pH value of about 3.5. As the majority of all analyses using LC/MS are carried out in the positive ionization mode, this approach is useful. However, the user should be aware that this is not a buffer system. Depending on how long the mobile phase remains in the solvent bottle and whether an exchange with the surrounding air takes place, a change in the pH value can occur which can be more or less noticeable. This, in turn, has a particularly strong effect on the retention behavior of those components with pK_a values close to the pH value of the mobile phase. Such a method is therefore not robust enough for these components. In the norm, it is specified that throughout six consecutive chromatograms the standard deviation in the retention time should not exceed 0.03 min. It is worth emphasizing that the standard deviation of the retention time is independant of the elution time of the compounds. With methods that have a total run time of only a few

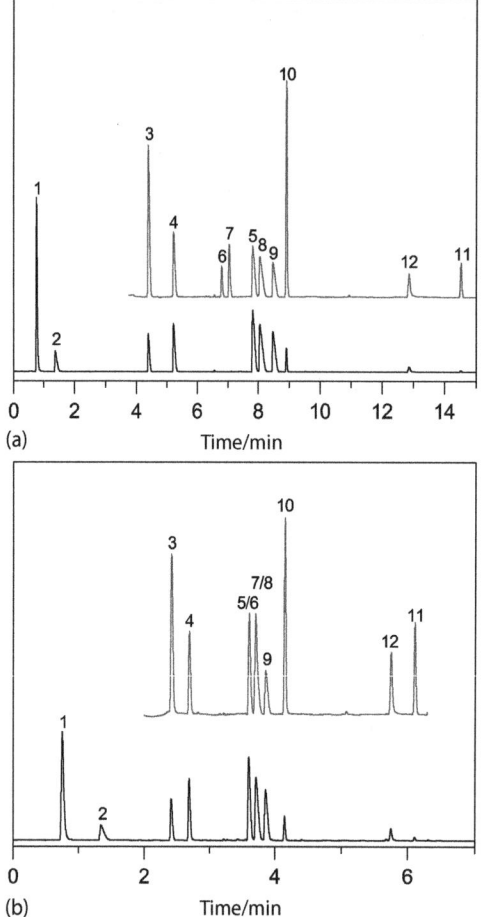

Figure 3.12 Comparative separation of 12 cytostatic drugs on an Agilent Zorbax SB C-18 (50 × 2.1 mm, 1.8 μm) column; chromatographic parameters: (a) solvent gradient: in 30 min from 1 to 99% B, (b) solvent gradient: in 10 min from 1 to 99% B; temperature: 30 °C; injection volume: 5 μL; mobile phase: A = water + 0.1% formic acid, B = acetonitrile + 0.1% formic acid; flow rate: 350 μL min^{-1}; detection: UV at 200 nm and 254 nm.

minutes, a deviation of 0.03 min is of course much more critical to judge than one with a run time of 30 min. Although the pH value exerts a strong influence on the robustness of the method, this parameter is often completely neglected in many fields of LC/MS analysis. Here again is a fundamental difference between labs in the pharmaceutical industry and environmental labs that only use LC/MS. In the norm, a reference is simply made to the addition of 0.1% acetic acid and the problem of inadequate pH value control is thereby ignored. As, in contrast to partially unspecific UV detection, better peak tracking is possible because of the characteristic mass transitions, pH value dependent fluctuations in the retention

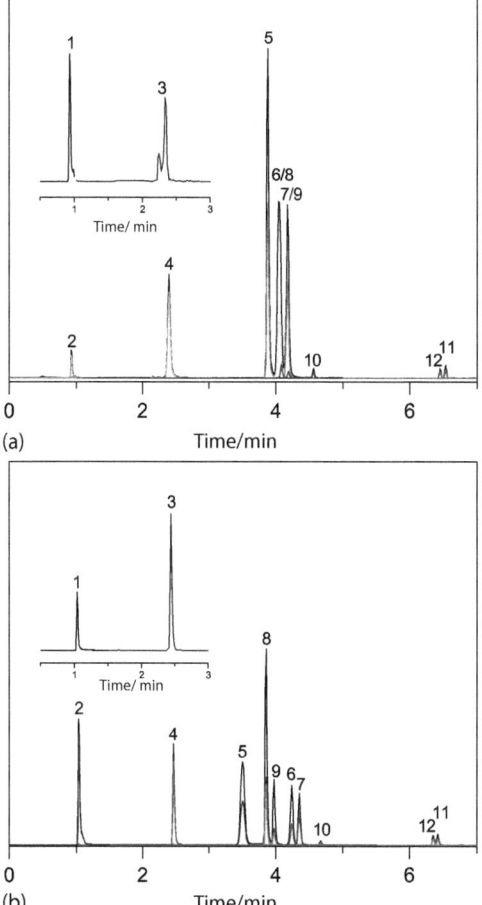

Figure 3.13 Comparative separation of 12 cytostatic drugs on a ChromaNik Sun-Shell RP-Aqua C-28 (50 × 2.1 mm, 2.6 µm) column; chromatographic parameters: (a) temperature: 40 °C; injection volume: 100 µL; mobile phase: A = water + 0.1% acetic acid, B = acetonitrile + 0.1% acetic acid; flow rate: 350 µL min^{-1}; (b) temperature: 40 °C; injection volume: 100 µL; mobile phase: A = water + 0.1% formic acid, B = acetonitrile + 0.1% formic acid; flow rate: 350 µL min^{-1}; detection: MS – multiple reaction monitoring.

time are often knowingly accepted. Using the following example, we would like to emphasize the influence that adding acetic and formic acid has on the separation of the substances listed in Table 3.2.

The separation presented in Figure 3.13a shows that a double peak is obtained for methotrexate (3) when the pH value is adjusted using 0.1% acetic acid. The reason for this is that the pK_a value of methotrexate lies at 3.41 and by adding acetic acid, the resulting pH value lies very close to the pK_a value. The consequence is a dissociated and nondissociated species, which can be separated on the column under the given conditions. In contrast, a symmetrical peak can be observed if 0.1% formic acid is added to the mobile phase (Figure 3.13b). The pH value is below 3 and, therefore, further apart from the pK_a value of methotrexate, so that only one species is present. The more complex the method, that is, the higher the number of target compounds to be recorded during one chromatographic run, the less probable it is that optimal chromatographic conditions will be achieved for all compounds. A generic LC/MS method is, therefore, always a compromise between the best possible chromatographic resolution and the special and often limiting technical specifications of mass spectrometry. On comparing the chromatograms in Figure 3.13, it is evident that the selectivity and elution order are different. Adding 0.1% formic acid results in a significantly better separation of the target compounds.

At this point we would like to refer the reader to further literature on how targeted optimization of the method can be carried out taking the pH value and buffer systems into consideration. The group of Roses and Bosch has published numerous papers on this and described in detail the influence of organic solvents on the changes in the pH value in the water and organic phase [29]. For practitioners, we direct you to the website of the University of Liverpool which contains a buffer calculator [30]. The user has the opportunity to define important factors for setting up the buffer using an interactive input field. If possible, the software then calculates the appropriate "recipe" for the correct buffer preparation. The temperature at the preparation of the buffer and the temperature specified for the chromatographic method can both be stipulated here. The book of the website's author is also worth reading. It lists briefly and concisely everything worth knowing about the topic of pH values and buffers for practical application in the lab [31]. This contains an extensive attachment of all buffer systems that are suitable for application with mass spectrometry. Further practical information on the importance of the pH value for chromatography as well as for the optimization strategy can be found in the book *HPLC Made to Measure* [32].

3.3.7
Using Simulation Software for Fine Optimization

If the screening experiments were not satisfactory, fine optimization has to be carried out with special simulation software. Examples of such software packages are Drylab, which was further developed by the Molnár Institute for Applied Chromatography or ChromSword, developed by Dr. Galushko. We are not going to explain in detail here how these software tools work. The interested reader can find an excellent, short description in Chapter 4 of the book *HPLC Made to Measure*. Instead, we would like to give a general description of how a method can be optimized using Drylab.

In this case, four gradient runs are entered into the software as basic data. One gradient with a low gradient slope should be applied and one with a higher slope so, for example, a gradient with a run time of 30 min and another of 10 min. As the temperature has a substantial influence on the selectivity, the corresponding gradient runs are included in the simulation with a low and high temperature, for example, 30 and 50 °C. In principle, this follows the approach of the experimental design presented in Figure 3.5. Using this matrix, the software can generate a resolution map. Using the basic data, it is possible to simulate chromatograms with varying gradient slopes and temperatures. A more complex gradient run that includes isocratic steps, for example, often leads to a better resolution of critical peak pairs but at the same time to broader peaks. Although the chromatographic resolution can be improved in this way, the peak broadening has an unfavorable effect on the signal-to-noise ratio. This is a clear disadvantage, especially when low detection limits have to be achieved. Therefore, it would probably make sense to carry out the elution using a simple linear gradient in order to obtain an adequate chromatographic resolution of critical peak pairs, while simultaneously optimizing the peak width so that an acceptable signal-to-noise ratio is achieved for the lowest concentration that is to be detected. Thus, our recommendation would be to carry out the elution using a linear solvent gradient without isocratic steps. This offers advantages in method transfer to other HPLC systems with different system and dwell volumes.

At this point, it is also critical to note that this kind of chromatographic optimization receives either little or no consideration in many fields of LC/MS analytisis. This fact probably arises from the misjudgment mentioned at the beginning that chromatographic separation is not really necessary when using a mass spectrometer. Besides that, it is not possible to completely separate several hundred components in one run. Current methods in the field of pesticide screenings contain up to 400 target compounds. Isobaric species that cannot be chromatographically resolved have to be quantified as sum parameters. The intelligent application of chromatographic simulation software would still prove advantageous if a connection to mass spectrometric parameters were possible. It has been noticed that providers of LC simulation software are reacting to this trend and are considering these aspects in their new software versions. We, therefore, encourage those users who are considering using or purchasing such software to directly ask the software vendors about specific criteria. Due to the fast development in this field, any information in this book about the current status of software programs would already be out of date at the time of printing and so there will be no further discussion of the topic here.

3.3.8
Choosing the Stationary Phase Support

Even though the selectivity of the phase system is without doubt the most important parameter in developing a chromatographic method, it has a different position with regard to mass spectrometry compared to UV detection. In particular, with generic LC/MS methods for recording a high number of components (> 50), other criteria such as the size of the particles of the stationary phase or the general question of the appropriate chromatographic support are also important.

Currently, three different chromatographic supports can be distinguished: fully porous particles with a diameter between 1.5 and 5.0 µm, core–shell particles with a thin porous layer on a solid core and a diameter between 1.3 and 5.0 µm, plus monolithic phases [33, 34].

The monolithic phases are characterized by a bimodal pore structure and are not made up of single particles. As they have relatively large flow-through pores, the permeability is much higher compared to particulate materials. This begs the question which support is suitable for which application in the context of LC/MS coupling?

During recent years, fully porous particles with a diameter of < 2 µm (so-called sub-2-µm particles) are widely used. A decisive advantage of these sub-2-µm particles is that the analysis time can be shortened significantly compared to using fully porous particles with a diameter of > 3 µm, when the same chromatographic resolution is obtained. The most important prerequisite here is an HPLC system whose extracolumn volume is as small as possible. Otherwise, the band broadening outside the column leads to a significant loss of efficiency. In LC/MS coupling, this means, among other things, the need for a transfer capillary between the column outlet and the inlet of the mass spectrometer which is often too long. Often, in the coupling of (U)HPLC and MS it is not possible to choose the optimum conditions for chromatography because of technical issues with the devices. For example, special capillaries are often built into the ion source which cannot be adapted in length and inner diameter. In addition, the individual modules of the (U)HPLC system and the mass spectrometer cannot always be aligned on a short distance. Figure 3.14 shows a typical example of how the devices are set up in many labs.

It is clear to see that the position of the (U)HPLC system with respect to the inlet of the mass spectrometer results in a more or less distinct system volume. In very unfavorable cases, the length of the transfer capillary after the column can be one meter. For the mass spectrometer shown in Figure 3.14, the ion source is located on the right. If the HPLC system is placed to the left, it is a relatively long way from the column outlet (outlet HPLC A) to the inlet of the ion source. If the HPLC system is set up on the right (HPLC B), the distance can be minimized. If the column is unfavorably placed in the HPLC oven, this can also lead to the need for a long transfer capillary. In addition, the capillary is often led through an extra switching valve on the mass spectrometer so that the part of the sample that has a high salt content is not directed into the MS. This is common practice when a large volume direct injection, for example, several hundred microliters is carried out. If the salts contained in the sample were deposited in the ion source or in the entrance of the ion path, this would lead to a continual reduction of the ionization efficiency after only a few injections. As can be seen by the labeling of the switching valve in Figure 3.14, the extra distance from the column outlet to the waste valve and from there to the inlet source is at least half a meter.

To minimize the band broadening after the column, one can adapt the inner diameter of the capillary. A smaller inner diameter (ID) inevitably leads to a higher pressure drop and will be responsible for a significant part of the total pressure

Figure 3.14 Typical system set-up in LC/MS coupling.

if the capillary is very long and has an ID < 50 μm. Depending on the type of ion source, there are system volumes that cannot be reduced. In fact, the influence of these system volumes plays a decisive role when the full separation performance of a highly efficient UHPLC column is to be taken advantage of. In many cases, a long transfer capillary with an inner diameter of 130 μm, for example, leads to a significant loss of the separation efficiency. The narrow bands generated in the column are partially or completely merged in the capillary as they leave the column. In addition, in the mobile phase prior to passing into the mass spectrometer, it is usual in some cases to add an extra "make-up" flow that contains ionization additives to improve the ionization rate of certain compounds. This is generally achieved using T-pieces and contributes to a substantial loss in the original separation efficiency. Therefore, it is worth considering whether stationary phases with a particle diameter of < 2 μm are appropriate for the device configuration available in the lab.

Things look very different in case it is possible to set up a direct connection between the column and the MS. In Figure 3.15, we show a system configuration where the flexible positioning of the HPLC system makes it possible to set up a connection between the column and the inlet of the mass spectrometer over a very short distance, thereby, significantly reducing the extracolumn volume.

In this case, the column oven bridges a large part of the distance which would otherwise have to be covered by a transfer capillary. In the meantime, many manufacturers have reacted to this problem and offer "intelligent" system solutions. With the aid of a flexible or freely adjustable HPLC oven, it is possible to min-

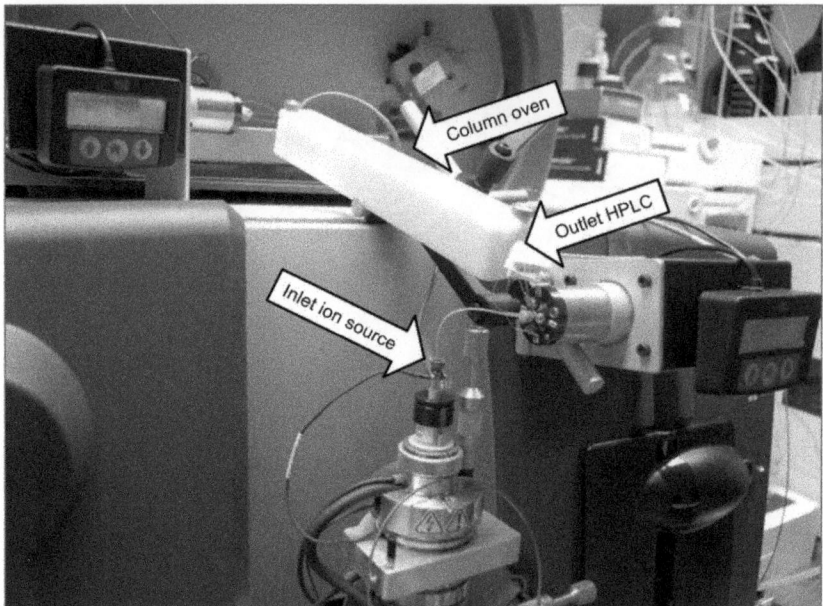

Figure 3.15 Optimized system configuration for coupling a flexible HPLC system with the mass spectrometer.

imize the critical system volumes after the column. However, the technical solutions specifically designed for optimum chromatographic conditions have less flexibility with regard to the choice of several columns. Column switching valves are very helpful during long measuring sequences over a weekend because different columns can be used for different methods. The column switching valves are usually directly integrated in the HPLC oven and, therefore, add to the overall dead volume. That is why a compromise again has to be accepted between the desired flexibility and the optimum chromatographic conditions.

As a general rule, one can say that when a system configuration as shown in Figure 3.15 is possible, and a mass spectrometer with fast scan and polarity switching times is used, the intrinsic efficiency of fully or partially porous sub-2-µm particles can be exploited to a very high degree. In other cases, monolithic phases represent an excellent alternative. The theoretically lower efficiency compared to the fully porous particles is not noticeable in practice because there is a lower separation efficiency per se due to the dispersion of the elution bands in the connecting capillaries. On top of this, in comparison to the particulate phases, there is a substantially lower total pressure under otherwise identical chromatographic conditions. This can be an advantage for reducing wear and tear on seals or, for example, when a coupling is set up with parts that are not designed for high pressure. In contrast to the classic stainless steel columns, monolithic columns are only available in a PEEK hardware. The manufacturer sets a pressure limit of 200 bar on these phases. When columns with a dimension of 50×2.1 mm are used, the pressure limit will not be exceeded – even with flow rates of up to 1 mL min^{-1} and with methanol as the organic solvent.

Besides these aspects, which focus exclusively on the extracolumn volume, the sample itself plays a decisive role regarding the choice of the appropriate particle diameter and the chromatographic support material. Analysis of relatively "clean" samples or inclusion of extensive sample preparation steps like purification and filtering direct to sub-2-µm particles as a good choice. Direct injection of a large volume makes phases with a particle diameter of about 3 µm or the already mentioned monolithic columns more suitable. A reason for this is the small interstitial volume in sub-2-µm phases which can cause blockages. Additionally, the inlet frits exhibit a lower mesh size than for 3 µm or even 5 µm particle packed columns. Monolithic phases have large flow through pores due to the bimodal pore structure so that colloidal compounds can be transported through the column without blockages occurring. For this reason, we have been using either monolithic stationary phases or fully porous phases with a particle diameter of 3 µm in our lab for the analysis of waste water samples.

3.3.9
The Influence of the Inner Column Diameter and the Mobile Phase Flow Rate

In order to guarantee fast cycle times and therefore high sample throughput, the inner diameter of the column should be chosen so that a high linear velocity can be achieved. In principle, when using a constant flow rate, the smaller the inner diameter of the column, the higher the linear velocity. This behavior is illustrated in Figure 3.16. The inner diameter of the column is plotted against the linear velocity of the mobile phase using a constant flow rate of 0.5 mL min^{-1}. The figure shows that the linear velocity increases significantly when the inner diameter is reduced and the flow rate is constant. For LC/MS, a flow rate of 0.5 mL min^{-1} represents a good compromise between sufficient ionization efficiency and fast analysis times.

Furthermore, this means that using a constant flow rate, columns with a small inner diameter are ideally suited to reduce the solvent consumption. The consistent miniaturization of separation systems is, therefore, the best method for minimizing both the analysis time and resource consumption. We will be returning to this point at the end of the chapter. Furthermore, the reduction of the inner diameter in combination with mass spectrometry is also worth aiming for because electrospray ionization with flow rates of < 0.5 mL min^{-1} often exhibit the highest efficiency. This is the reason why columns with an inner diameter of 2.1 mm as opposed to those with an inner diameter of 4.6 mm are meanwhile preferred in most labs that use mass spectrometry. Otherwise, using a column with an inner diameter of 4.6 mm reduces the influence of the system volume on the dispersion of the elution bands.

Using columns with an inner diameter of 2.1 mm, the dispersion volume after the column can lead to a lower chromatographic efficiency. This is similar to the problem described above when the particle diameter of the stationary phase is

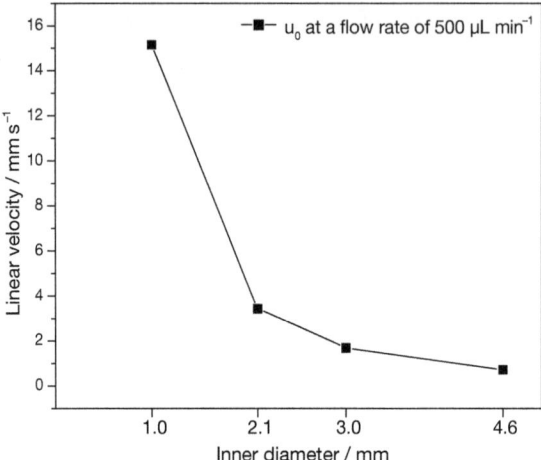

Figure 3.16 The inner diameter of the column plotted against the linear velocity of the mobile phase for a constant flow rate of 0.5 mL min^{-1}.

reduced. Therefore, we can say that for LC/MS coupling one needs to aim for a system configuration where the column is connected to the inlet source of the mass spectrometer via the shortest route.

3.3.10
The Influence of the Injection Volume

3.3.10.1 Direct Injection
It is often the case with multianalyte methods that global limits of detection (LOD) and limits of quantification (LOQ) are defined. From the point of view of a layperson, this approach may seem totally understandable; however, it does not take into account that no matter which detection technique is used, there is always an analyte-specific dependency on the detection limit. Without a chromophoric system, a substance is not UV active, which can be seen in Figure 3.17 where the UV spectra of cyclophosphamide and ifosfamide are presented. An analogue principle is valid for mass spectrometry. This is far away from being a universal detection method because ionization is also substance specific. For example, many compounds cannot be ionized using electrospray ionization. In these cases, one has to use other detection or ionization techniques. Nevertheless, many government agencies and industrial clients demand that a strictly defined detection limit, for example, 10 ng L^{-1} or even lower, should be the global goal for practically all compounds in a sample that are to be investigated. This is in fact possible in many cases but often with significantly more effort.

In environmental analysis, a technique that is known as large volume injection (LVI) has therefore become established [35, 36]. This involves injecting a relatively large volume compared to the dimensions of the column, for example, 1 mL, into columns with an inner diameter of 2.1 mm and 4.6 mm, respectively. This

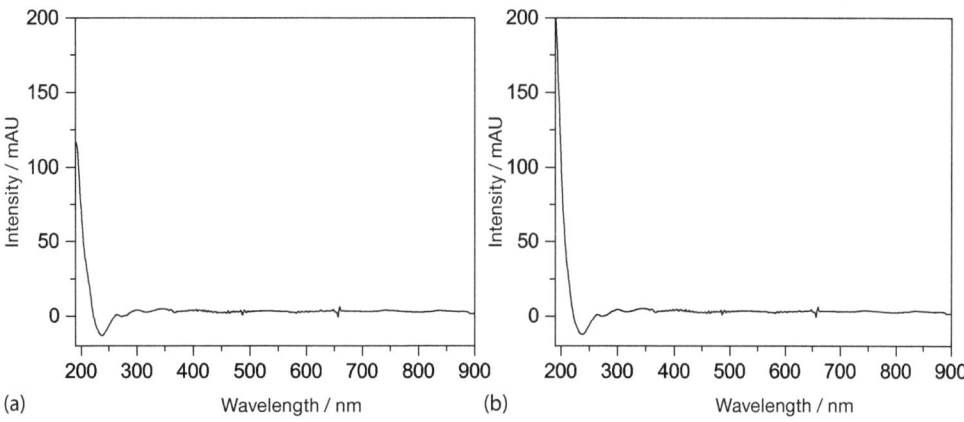

Figure 3.17 UV spectra of (a) cyclophosphamide and (b) ifosfamide.

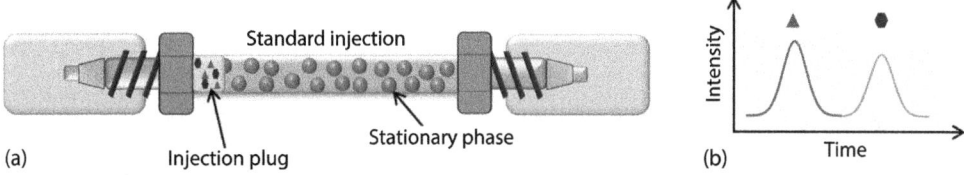

Figure 3.18 Schematic representation for the "normal" injection: (a) HPLC column, (b) chromatogram.

approach is very efficient from the analytic point of view because apart from a simple sample filtration there are no other steps necessary for determining the target compounds in low concentrations. Normally, the maximum injection volume should not exceed 10%, preferably 5%, of the column void volume. How is it then possible to inject up to 1 mL, which is several times more than the capacity of the column? Let us take a look at the normal injection which is presented here in Figure 3.18. Figure 3.18a represents a column, while Figure 3.18b illustrates the chromatogram with the corresponding elution profiles for a polar and nonpolar compound.

The injection band marked with a plug only takes up a small part of the whole column volume. Independent of whether the elution is isocratic or achieved using a solvent gradient, symmetrical peak forms will be obtained as long as there is no reason for tailing to occur. This is valid for both polar and nonpolar compounds. The composition of the injection solution has little or only a negligible influence on the peak shape.

Things are different when a large volume is injected directly onto the column. In the first case (Figure 3.19a), the sample is dissolved in water. The injection plug takes up about half of the volume available to the mobile phase in the column. At this moment, the HPLC column becomes a solid phase extraction column. Specifically, most of the compounds will be concentrated at the head of the col-

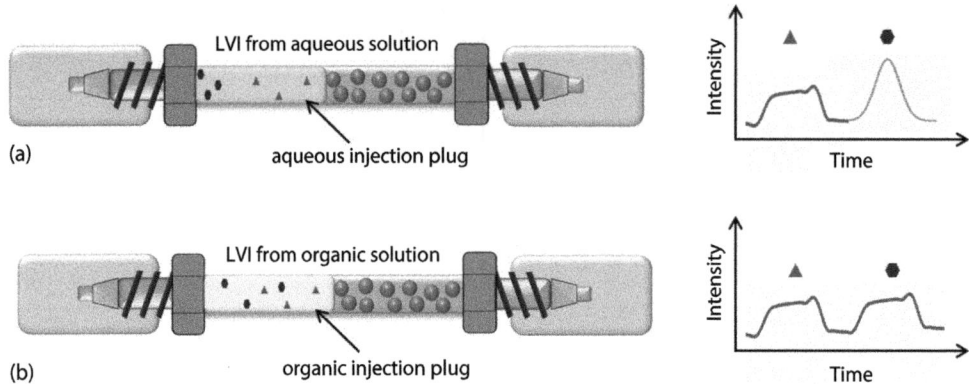

Figure 3.19 Schematic representation of a large volume direct injection for (a) injection from aqueous phase and (b) injection from organic phase. The injection plug takes up approximately half of the volume available to the mobile phase in the column.

umn because water is the weakest elution solvent in RP chromatography. As the interaction of the polar compounds with the stationary phase is very low, these components will have already been transported through the column with the injection plug. In the worst case, the elution bands of the polar portion cannot be compressed by the solvent gradient resulting in double peaks or strongly distorted or extremely broad peak profiles. The nonpolar compounds can still be eluted as symmetrical and narrow bands as no band broadening occurred through the large volume direct injection. This can be seen in Figure 3.19a where the corresponding chromatogram is presented on the right hand side. If the LVI is carried out using a 100% organic solvent, practically all substances will be transported through the column with the injection plug, so that for almost all compounds, double peaks can be observed. In reality, very different peak profiles form. In some cases the substance zones are "smeared" over a very wide area of the chromatogram, while for other substances there are clearly defined double peaks to be seen. As a general rule, one can say that a large volume injection will only makes sense if there is sufficient interaction between the analytes and the stationary phase.

Now we would like to look deeper into the theoretical background using a couple of selected examples from practice. It is helpful here to get an accurate picture of the volumes which we will be discussing next. First, we have to define what volume of the mobile phase in the column is really available. At first glance, this could be the same as the geometric volume of the column which can be calculated using the length and diameter according to the equation below:

$$V_{\text{Column}} = \frac{d^2_{\text{Column}}}{4} \cdot \pi \cdot L$$

As the stationary phase takes up a certain volume, this has to be subtracted from the geometric volume of the column. To keep things simple, let us assume that the porosity ϵ, which is a measure for the permeability of the stationary phase, is a constant factor of around 70% for silica-gel-based reversed phases. Accordingly,

Table 3.4 Comparative overview of the effective column volumes depending on the length and diameter of HPLC columns.

Inner diameter (mm)	2.1			3.0			4.6		
Length (mm)	50	100	150	50	100	150	50	100	150
$V_{\text{column,effective}}$ (µL)	121	243	364	247	495	742	582	1163	1745

the geometric column volume is simply multiplied by the porosity to obtain the available column volume of the mobile phase or to calculate the effective column volume.

$$V_{\text{Column,effective}} = \frac{d^2_{\text{Column}}}{4} \cdot \pi \cdot L \cdot \varepsilon$$

Table 3.4 presents the available volume of the mobile phase for the most common inner diameters and lengths. As the inner diameter becomes smaller the column volume falls likewise.

As already mentioned, in LC/MS coupling short columns of 5 cm in length with an inner diameter of 2.1 mm are usually used. Therefore, a direct injection of 1 mL means that the injection plug is around 8.3 times the size of the column volume. In other words, the column will be flushed with the injection solvent more than eight times before the gradient separation actually starts. Let us take a look at concrete examples from real life. In the first example, a method should be developed for the compounds listed in Table 3.2 that facilitates the determination of all components in a concentration of up to $0.1\,\text{ng}\,\text{mL}^{-1}$. The

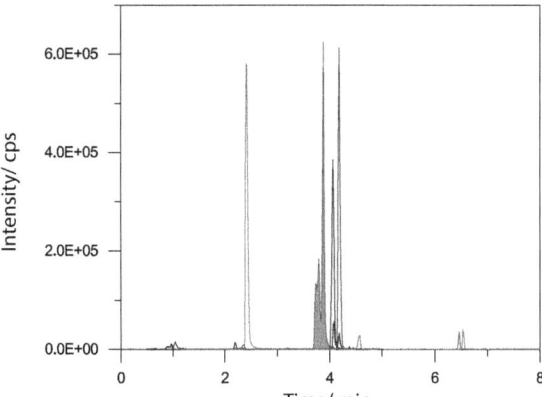

Figure 3.20 Separation of 12 cytostatic drugs on a ChromaNik SunShell RP-Aqua C-28 (50 × 2.1 mm, 2.6 µm) column; chromatographic parameters: temperature: 40 °C; injection volume: 5 µL; mobile phase: A = water + 0.1% acetic acid, B = acetonitrile + 0.1% acetic acid; flow rate: 350 µL min^{-1}; detection: MS – multiple reaction monitoring; composition of the injection solution: 30/70 (v/v) water/isopropanol. The dark area is the elution band of irinotecan.

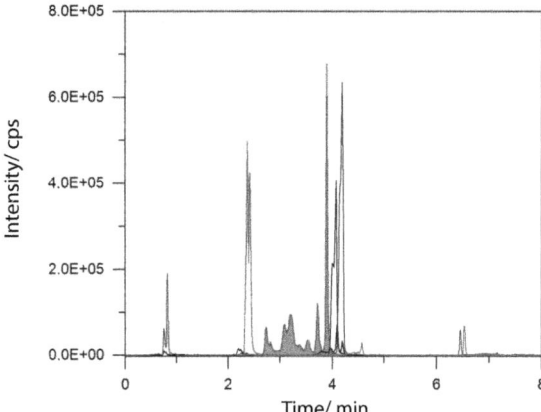

Figure 3.21 Separation of 12 cytostatic drugs on a ChromaNik SunShell RP-Aqua C-28 (50 × 2.1 mm, 2.6 µm) column; chromatographic parameters: temperature: 40 °C; injection volume: 10 µL; mobile phase: A = water + 0.1% acetic acid, B = acetonitrile + 0.1% acetic acid; flow rate: 350 µL min^{-1}; detection: MS – multiple reaction monitoring; composition of the injection solution: 30/70 (v/v) water/isopropanol. The dark area is the elution band of irinotecan.

chromatogram shown in Figure 3.20 was obtained after injecting 5 µL of reference standard that contained 10 ng mL^{-1} per analyte. The dark area marks the elution band of irinotecan. This is not an example of several components that have partially co-eluted, but it is one single compound that is smeared across a broad area.

The injection solution contained 70% isopropanol and the column volume available for the mobile phase was 121 µL. It is therefore surprising that no symmetrical peak form for all compounds was obtained. From pure intuition, one could have expected that the first components to elute would have been particularly affected by a peak deformation. But for these substances, as with all other analytes, symmetric peak forms were obtained. In a second step, an attempt to increase the injection volume from 5 to 10 µL was made as, for the compounds eluting at the beginning of the chromatogram, the signal intensity is too low (Figure 3.21). Now there are several peaks smeared across a large area although the injection volume is still less than 10% of the effective column volume.

The strategy of increasing the injection volume to enhance the signal intensity did not lead to success even though it could be increased significantly for the components that eluted first. The only way out of this dilemma and to achieve the detection limit of 0.1 ng mL^{-1} for all compounds is to dilute the sample solution with water by a factor of 10 and increase the injection volume subsequently by the same factor. Although this approach does not seem intuitively logical, the chromatogram presented in Figure 3.22 is convincing evidence that the assumption is correct. Why did diluting the sample by the same factor as the subsequent increase in the injection volume lead to the solution of the problem?

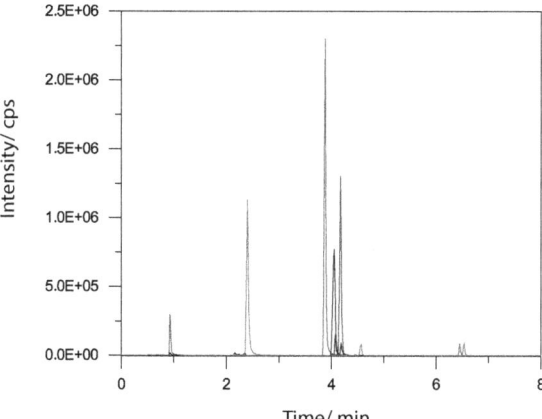

Figure 3.22 Separation of 12 cytostatic drugs on a ChromaNik SunShell RP-Aqua C-28 (50 x 2,1 mm, 2,6 μm) column; chromatographic parameters: temperature: 40 °C; injection volume: 100 μL; mobile phase: A = water + 0.1% acetic acid, B = acetonitrile + 0.1% acetic acid; flow rate: 350 μL min^{-1}; detection: MS – Multiple Reaction Monitoring; composition of the injection solution: 93/7 (v/v) water/isopropanol.

Originally, the injection solution contained 70% isopropanol. After dilution, this amount was reduced to only 7%. Despite the fact that under the given conditions about 82.5% of the column volume was occupied by the injection plug with a volume of 100 μL, it is quite clear that sufficient focusing for all substances was achieved. How can this be? In RP chromatography, there is an empirical relationship between the natural logarithm of the retention factor (ln k) and the proportion of the organic solvent in the mobile phase (% B). This functional relationship is the basis for the so-called linear solvent strength (LSS) model that is used in RP chromatography to predict the retention on the basis of two gradient runs [37]. The higher the proportion of the organic solvent in the mobile phase, the lower the corresponding retention of a compound on a reversed phase column. Indeed, isopropanol is a significantly stronger eluent than, for example, methanol or acetonitrile. Even if only a small volume in relation to the effective column volume is injected, the high proportion of isopropanol in the injection solvent prevents the component to concentrate at the column head, whereby a certain extent has already been transported through the stationary phase with the injection solvent. This effect can be so strong that the elution band can no longer be compressed in spite of a gradient elution and the result is a strongly distorted peak profile. Diluting the sample with water has a counter effect so that the injection volume can even be increased substantially. With the help of this strategy, the aim of a global detection limit for all substances of 0.1 ng mL^{-1} can be achieved.

The insufficient retention of the polar molecules such as 5-fluorouracil or gemcitabine is a problem that has still not been solved. All the phase systems described in Sections 3.3.6.1 and 3.3.6.2 were not suitable for retaining these components with a retention factor of > 2. Here, a different phase material is needed such as porous graphitic carbon (PGC) better known as Hypercarb.

Using this phase material, it is also possible to retain very polar substances [38]. Here, it is important to remember that nonpolar substances can also be strongly retained and there is the risk that these components will be irreversibly bound or elute from the column very late. Hence, we will now describe the injection of large volumes for the enrichment of polar and nonpolar substances on a coupled phase system. A short PGC precolumn (10 × 2.1 mm, 5 µm) was serially connected to a classic C-18 reversed phase (50 × 2.1 mm, 3.5 µm). The assumed advantage of using this column combination is that with the aid of the Hypercarb precolumn, sufficient retention of strongly polar components will be achieved and, furthermore, the retention for nonpolar substances is not so strongly pronounced. First, we are going to show the influence of the Hypercarb precolumn on the retention

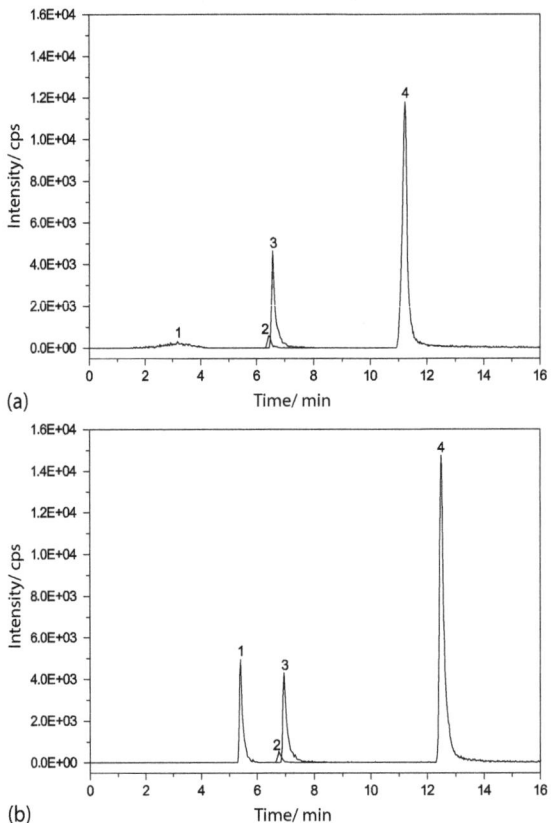

Figure 3.23 Separation of four cytostatic drugs (a) without and (b) with online enrichment. Chromatographic conditions: stationary phase: (a) Waters XBridge C-18 (50 × 2.1 mm, 3.5 µm), (b) Thermo Hypercarb (10 × 2.1 mm, 5 µm) coupled with a Waters XBridge C-18 (50 × 2.1 mm, 3.5 µm) column; mobile phases: A) water with 0.1% trifluoroacetic acid, B) acetonitrile with 0.1% trifluoroacetic acid; flow rate: 0.5 mL min^{-1}; gradient: 0 to 90% B in 10 min, 10–20 min, 100% B; injection volume: 1000 µL; temperature: 35 °C; detection: MS. analytes: 1) gemcitabine, 2) ifosfamide, 3) cyclophosphamide, 4) fenofibrate.

of polar, medium polar, and nonpolar analytes. In Figure 3.23, there are two chromatograms which present the separation of four pharmaceutical drugs.

Figure 3.23a shows the chromatogram of the separation without the Hypercarb precolumn, therefore without the online enrichment, whereas in Figure 3.23b the separation with the PGC precolumn is presented. A comparison of the two chromatograms emphasizes that just as with the nonpolar fenofibrate, for the medium polar ifosamide and cyclophosphamide there is no difference regarding the peak width. The injection of an aqueous standard of 1000 µL containing these analytes can be carried out without having a negative influence on the peak shape. In comparison, the polar gemcitabine elutes from the C-18 column as a broad unfocussed band if the precolumn is not used. The retention of gemcitabine on the C-18 material is so low that enrichment is impossible under these conditions. However, a combination of a PGC precolumn and a C-18 reversed phase (Figure 3.23b), leads to a focused gemcitabine peakat the head of the column and can be subsequently eluted as a narrow band with the solvent gradient. With regard to the optimum injection volume of $\leq 10\%$ of the column volume (≤ 12 µL), an enrichment factor of 83 is obtained.

Finally, it should be shown that the injection of larger volumes does not necessarily result in a significant loss in chromatographic efficiency. In Figure 3.24, the separation of the pharmaceuticals is compared with regard to injection volumes of 5 and 1000 µL. The differences in the retention times can be explained by the

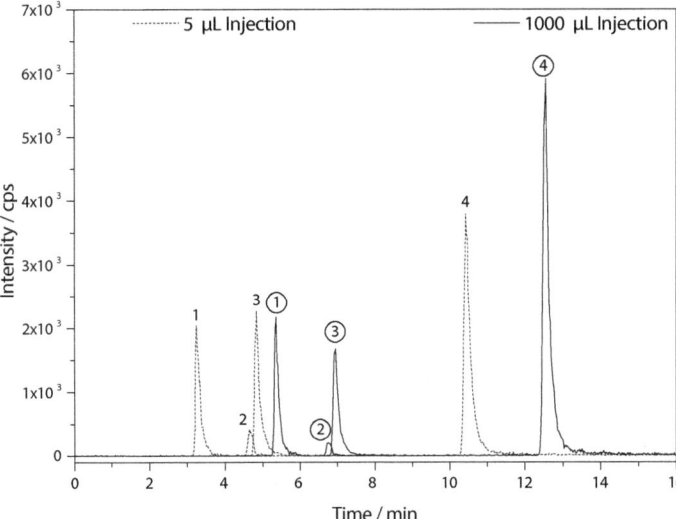

Figure 3.24 Separation of four cytostatic drugs. Chromatographic conditions: stationary phase: Thermo Hypercarb (10 × 2.1 mm, 5 µm) coupled with a Waters XBridge C-18 (50 × 2.1 mm, 3.5 µm) column; mobile phase: A) Water with 0.1% trifluoroacetic acid, B) acetonitrile with 0.1% trifluoroacetic acid; flow rate: 0.5 mL min^{-1}; gradient: 0 to 90% B in 10 min, 10–20 min, 100% B; injection volume: see diagram; temperature: 35 °C; detection: MS. Absolute amount of substance on column: 25 ng. Analytes: 1) gemcitabine, 2) ifosfamide, 3) cyclophosphamide, 4) fenofibrate.

different volumes of the sample loops. For the injection of 5 µL, a sample loop with a volume of 5 µL was used, whereas for the injection of 1000 µL a loop with a volume of 1000 µL was used. The temporal offset of 2 min therefore corresponds with the flow time through the 1000 µL sample loop at a flow rate of 0.5 mL min^{-1}.

In both cases, the same absolute amount of substance, 25 ng per analyte, was applied to the column. The comparison of the peak shapes and the peak widths of the analytes shows that there are no negative effects caused by the injection of 1000 µL. An additional peak dispersion caused by the high injection volume could not be observed.

Direct large volume injection in combination with mass spectrometry is successfully used in the field of environmental analysis because no additional pumps or consumables for the sensitive determination of a large number of compounds in one single chromatographic run are necessary. Large volume injection always reaches its limits when the sample contains very hydrophobic species that can only be eluted from the column by using lengthy rinsing steps with a nonpolar solvent. Due to the high injection volume, there is a risk of a clogged column after only a few injections. In addition, the compounds that will elute continuously over a long period from the column lead to an increase in noise which can considerably deteriorate the signal-to-noise ratio. In this case, there is no alternative to sample clean-up and enrichment using online or offline SPE.

3.3.10.2 Online SPE

In the meantime, several manufacturers offer systems that allow direct enrichment in the coupling with HPLC – we call this online SPE. In contrast to offline SPE, online SPE offers a series of advantages. Below we introduce a method for the direct coupling of an online solid phase extraction with liquid chromatography and tandem mass spectrometry (online SPE-LC/MS/MS) which simplifies and speeds up the workflow in water and waste water analysis.

Figure 3.25 shows an online SPE-LC/MS/MS system that was used for method development. It is made up of a "Prep and Load Robotic Tool Change" (PAL RTC) autosampler with a 100 µL, a 1 mL, and a 10 mL syringe, two injection valves with different sample loops, and an enrichment unit for automated online SPE with a cartridge changing unit. The first sample loop has a volume of 50 µL so that the system can be used as a conventional HPLC system with the opportunity for

Figure 3.25 System configuration of the online SPE/LC/MS/MS coupling: 1) HPLC pumps, 2) PAL RTC autosampler, 3) automatic cartridge exchanger, 4) HPLC oven, 5) tandem mass spectrometer.

applying small sample volumes. The second sample loop has a volume of 10 mL. This is used for enriching the sample on an exchangeable SPE cartridge. Rinsing steps can be carried out for removing salts. The transfer of the sample from the online SPE cartridge to the chromatographic column occurs in this case with the flow of the solvent gradient which is also used for the HPLC separation. As can be seen when comparing the two chromatograms in Figure 3.26, the peaks that were enriched on a polymer material (resin SH) elute as broad bands. Compared with this, the components that were enriched on a C-18 cartridge can be detected as narrow peaks.

One possible explanation for this could be the large particle diameter of 20 to 50 µm for the material of the resin SH cartridge, as opposed to 7 µm for the material of the C-18 cartridge. A way to circumvent this problem is by eluting the enriched sample as a plug with purely organic solvent from the SPE cartridge. This does mean, however, that an extra pump would be needed to dilute the plug with water before it reaches the HPLC column (see the explanations in Section 3.3.10.1). Otherwise, it would not be possible to focus the compounds present

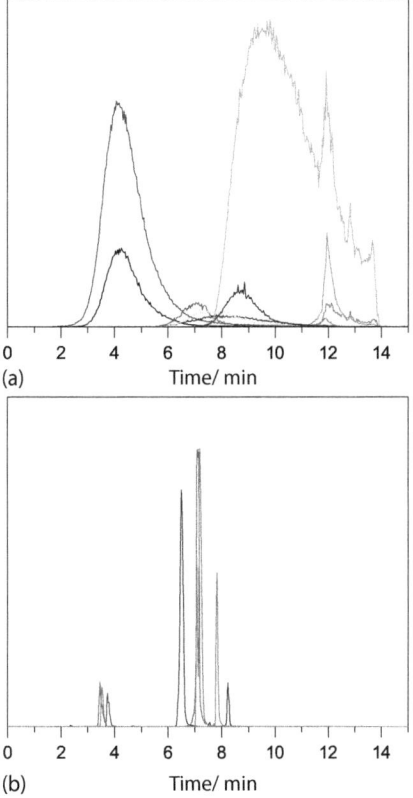

Figure 3.26 Comparison of the chromatograms of an enriched sample on (a) resin SH cartridge and (b) C-18 cartridge.

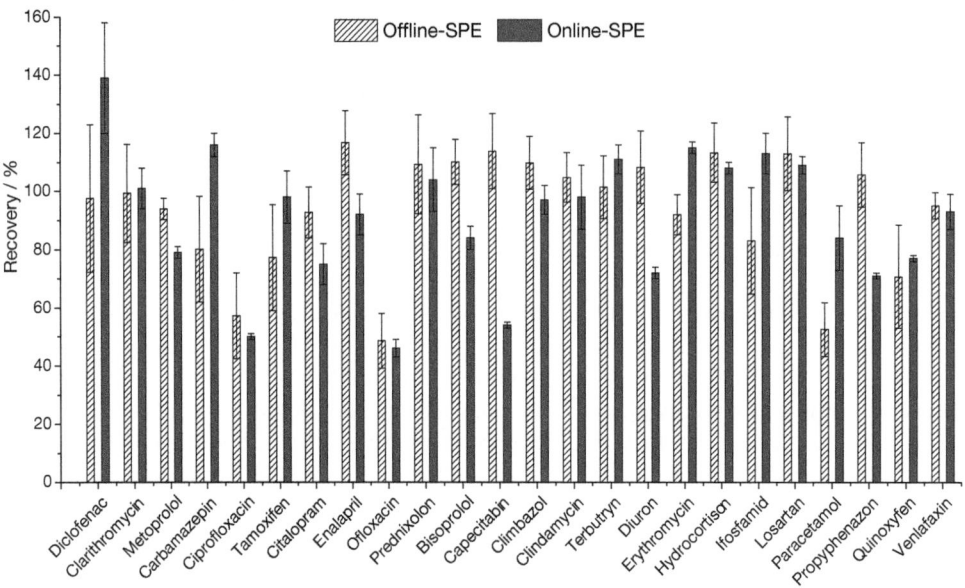

Figure 3.27 Comparison of the recovery rates of online SPE (C-18 cartridge, $n = 3$) and offline SPE (Waters HLB cartridge, $n = 5$) for a waste water-multianalyte method with 25 pharmaceuticals and an injection volume of 10 mL in the online SPE.

in the organic solvent and the result would be strongly distorted peaks. A limitation of the cartridge changer used here is a maximum pressure of 300 bar, which should not be exceeded as leaks could occur. For this purpose, it would make sense to use a monolithic column which would give a pressure of less than 200 bar using a flow rate of 0.5 mL min^{-1}, even when methanol is used as the organic solvent (see explanations in section 1.3.8). Figure 3.27 shows a comparison of the recovery rates for selected pharmaceuticals that were enriched using online and offline SPE. For most compounds no significant differences can be observed.

The final point concerns the interlacing of all the steps. Figure 3.28 is a good comparison of the time scheduling when samples are analyzed with online and offline SPE. For example, 1 h is needed for the manual preparation of a single sample. This does not include the measurement time. Using online SPE, the time can be reduced by about half as all manual steps fall away. With the help of the software CHRONOS (Axel Semrau, Sprockhövel, Germany), the steps can be interlaced so that the enrichment of one sample can take place while another sample is analyzed. The whole analysis time can then be reduced by half again.

Online SPE with its complete automation is significantly superior to offline SPE with regard to measurement uncertainty and time saving, although it can be noted that with the appropriate equipment, a lab can enrich several samples simultaneously using offline SPE and the time advantage of online SPE therefore shrinks. It must also be acknowledged that the manual intervention of the lab personnel is reduced to a minimum and by using the night-time and weekend sequencing

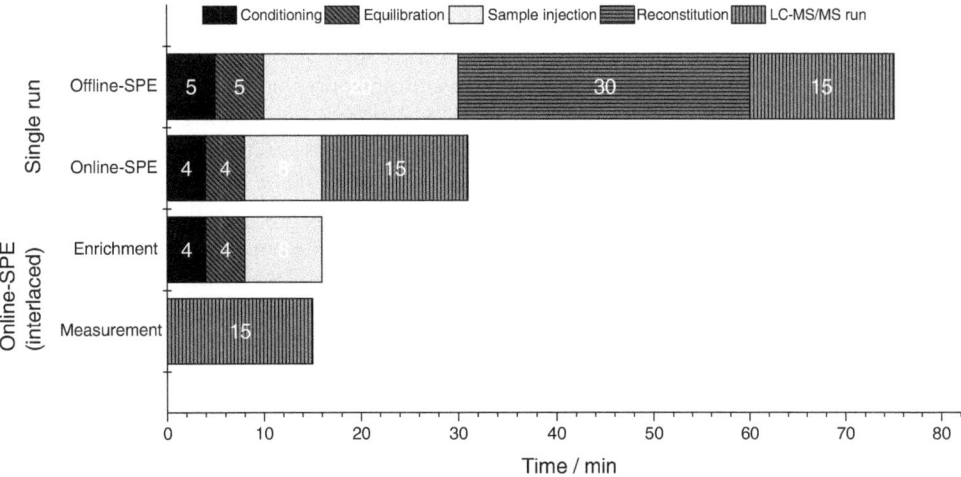

Figure 3.28 Time schedule for sample preparation in manual offline SPE and the automated online SPE.

the work on many samples can be automated. Considering the continually rising costs of personnel, online SPE is a real alternative to the established and time-consuming manual sample preparation.

3.3.10.3 Comparative Overview of the Enrichment Methods

Direct injection of large volumes can be used successfully when relatively "clean" samples such as surface or drinking water are to be analyzed. In order to avoid undesired salt loads from entering the ion source of the mass spectrometer, switching valves need to be used wherever possible to direct these sample components to the waste. An alternative to large volume direct injection or offline SPE is to use online SPE. The lifetime of the analytical column can be significantly increased by using the SPE cartridge to retain strongly hydrophobic compounds that are contained in the matrix. Regarding the analysis of polar compounds, stationary phases based on porous graphitic carbon have proved reliable. These permit sufficient retention in relation to the void time so that the quantification is not compromised by co-eluting salts.

3.3.11
Establishing the Mass Spectrometric Parameters

3.3.11.1 Introduction

Next, we would like to discuss the strategy of optimizing the mass spectrometric parameters to obtain the highest possible signal intensity for all compounds to be determined in one method. In the norm it says, *Following the defined conditions for chromatography the optimum setting for ionization should be chosen for every substance in the positive or negative mode taking the chemical characteristics of the substance into consideration.* If only two to five components have to be analyzed in a method, this is definitely possible as the mass spectrometric parameters can, to some extent, be switched during the chromatographic run. The more substances to be recorded in one run, the more important it is to define generic parameters. Consequently, it is essential to first define the mass spectrometric parameters that are most important. We would like to emphasize that the parameters described in the following sections relate to a mass spectrometer from the company SCIEX. Besides the terminology that is used, this also applies to the recommendations for certain settings such as the ionization voltage, etc. Depending on the device that is available, it could be necessary to include other parameters in the optimization. Nevertheless, the general approach is quite valid.

3.3.11.2 Orientation of the Sprayer Position

The position of the ESI needle within the ion source can have a significant influence on the signal intensity. At a low flow rate of $< 100\,\mu L\,min^{-1}$, for example, it is recommended that the ESI needle should be oriented close to the opening of the curtain plate. With higher flow rates, the distance between the ESI needle and the curtain plate should be increased. This is also the case for samples with very high matrix loads so that the entrance of the ion path does not become too dirty. The optimization of the sprayer position can be done using flow injection analysis of the analytes. In order to check the change in intensity, it is useful to record the mass transitions of all relevant substances over a certain time period in the MS software. Thereby, one can react to analyte-specific characteristics. If there are analytes in the method that show a reduced response, the needle can be optimized to the highest possible signal intensities or the highest signal-to-noise ratio for these compounds.

3.3.11.3 Curtain Gas

As already explained in Chapter 1, ESI is a technology that works at atmospheric pressure. In contrast, the mass spectrometric experiments are carried out in a high vacuum. The task now is to keep these parts separate or to avoid flooding the vacuum. This is ensured in LC/MS coupling by using the so called curtain gas (CUR). Usually, a nitrogen stream is used, which can be set by the user. In addition, the nitrogen curtain has the task of keeping nonionized neutral particles and matrix components away from the high vacuum area of the mass spectrometer to avoid contamination of the ion path which is inaccessible for the user. It is necessary here to ensure that a "useful" pressure is used for the curtain gas. If it is too low, this will lead to contamination of the device. If it is too high, the entry of the analyte will be negatively influenced in certain circumstances. In the process, it can happen that compromises are made between the "desired" degree of contamination and a higher entry of the analyte into the mass spectrometer. As a rule, it should be noted that the pressure for the curtain gas should only be high enough to have no negative influence on the intensity of the analyte. Nevertheless, every manufacturer has its own standard values for each mass spectrometer as, depending on the source design, peculiarities can appear that need to be taken into consideration.

3.3.11.4 Ionization Voltage

The ionization voltage means the electric potential that is applied to ionize the analytes. For ESI in the positive ionization mode, the voltage is generally between 3.5 and 5.5 kV. In negative ionization mode, the voltage is usually between −3.5 and −4.5 kV. As already mentioned above, these values are to be understood as guidelines that can vary depending on the system and substance. An optimization of the parameters is mandatory as they have a direct influence on the signal intensity. In extreme cases, an excessively high ESI voltage can lead to a bright glow discharge at the tip of the ESI needle inducing in-source fragmentation. Besides a disruption of ion transition into the mass spectrometer, voltage flashovers can occur in the ion focusing modules.

3.3.11.5 Source Temperature

The temperature setting has a direct influence on the evaporation of the mobile phase, but also on the analyte itself. This is why the temperature has to be set in a way that no undesired reactions as e.g. in source fragmentation can take place in the source. For thermosensitive and thermostable substances, a compromise has to be made regarding the optimum source temperature. In this case, the sensitivity of the individual substances can be optimized by regulating the gas stream. This will be explained in more detail below. However, if a higher number of components (> 15) is to be analyzed in one chromatographic run, a generic temperature needs to be chosen to guarantee a comprehensive analysis. Changing the source temperature during a chromatographic run does not make any sense because temperature equilibration needs several minutes.

3.3.11.6 Gas Flows

A distinction is generally made between nebulizer and dryer gas. The terminology can vary within the software, depending on the manufacturer. The nebulizer gas is the gas stream that passes the ESI needle and has an influence on the spray formation. An important thing to note here is that the pressure setting for the nebulizer gas is closely associated with the position of the ESI needle. The drying gas is used to ensure a homogeneous temperature distribution within the source. A low gas flow rate leads to poor temperature distribution. A higher flow rate leads to a better temperature distribution.

3.3.12
Optimization of the Mass Spectrometric Parameters

3.3.12.1 Introduction

The source parameters of the mass spectrometer are often underestimated parameters, although the chosen settings can have a significant influence on the resulting signal intensity. In general, the optimization should be guided by the number of analytes to be investigated. If a limited number of components is to be analyzed (≤ 15), it makes sense to do an analyte-specific optimization. However, for a multicomponent method, the optimization needs to be generic. As a basis for mass spectrometry optimization, a standard or reference solution is prepared which includes all analytes. Here, one needs to pay attention to the concentration. If it was too low, the signal might be too low. If it is too high, the signal could saturate the detector which means that a qualified statement about the optimum conditions would not be possible. Depending on the flow rate, a flow injection analysis (FIA) is performed using either a syringe pump or the autosampler. In principle, the optimization using FIA can be carried out in different ways as is described in the following section.

3.3.12.2 Manual or Partly Automated Optimization

Meanwhile, many instrument manufacturers support the user in the optimization of the source parameters with software implemented workflows. In the process, the user creates an LC/MS method that includes a chromatographic method, the specific mass transitions, and a preselection of appropriate source parameters. In the next step, the column for the flow injection analysis is replaced by a zero dead volume connector. In the optimization mode, one can later apply the created method. In general, the preferred chromatographic method is an isocratic eluent made up of 50/50 (v/v) water and organic solvent. Where possible, the flow rate should be similar to that of the actual analysis. It should be noted that the composition of the mobile phase also has a large influence on the ionization efficiency. If the analytes in the final method elute with a high proportion of water, then experiments should be carried out using a composition of 90/10 (v/v) water and organic solvent.

The different settings for each of the source parameters can be entered in the software. The software calculates the number of injections, including the required sample volume. In the process, one needs to make sure that there is enough volume of the standard. Once all settings have been defined, the FIA can be started. If all the settings are adjusted correctly, the system will autonomously go through all the chosen conditions one after the other. The autosampler of the LC system is used for the injection. At the end, the user is provided a report with the intensities of all mass transitions under the chosen conditions. Furthermore, the software creates a method with the appropriate parameters on the basis of the sum of all intensities. This kind of optimization is certainly more comfortable than classic manual optimization. Nevertheless, the user should check the resulting data carefully as in some cases peculiarities do occur. We will discuss these in more depth in Section 3.3.12.4.

3.3.12.3 Optimization Using Design of Experiments

Whenever a systematic evaluation of different parameters is to be carried out, statistical approaches such as design of experiments (DoE) or quality by design (QbD) are useful. There are some software packages that support the user. Examples of such software packages are Unscrambler or Fusion QbD. With the aid of these software tools, one can create experimental designs to investigate the influence of the individual parameters or mutual interdependencies on the signal intensity.

However, users do need to consider in advance what the most useful parameters are and which levels should be investigated. The generated experimental design can then be worked through step by step. Afterward, the targeted values (intensity or peak area) are put into the program. The software then calculates a global optimum where the maximum target size should be.

Such software packages are also useful with regard to the LC/MS coupling, as the manual optimization is often based on trial and error. To create an experimental design, the source parameters can be entered as variables. Ideally, the experiments should be carried out randomly.

3.3.12.4 Comparison of the Optimization Strategies

From Sections 3.3.12.2 and 3.3.12.3, it appears that optimization of the mass spectrometric source parameters is possible in different ways. Now the question is – which approach leads to better results? For evaluation, this will be discussed in the following, on the basis of the optimization of the signal intensity for 12 selected cytostatic drugs. The results of both approaches will be directly compared. The above-mentioned process-specific steps were carried out for both methods. Figure 3.29 shows a bar graph of the resulting intensities in percentages which are normalized to the maximum intensities of each analyte for the source tempera-

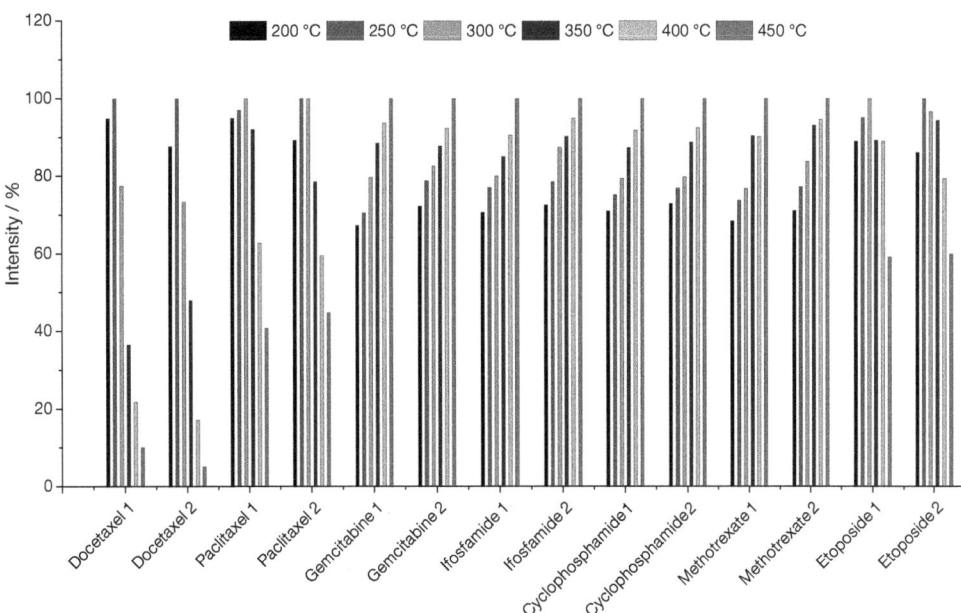

Figure 3.29 Resulting intensities for the two mass transitions of each analyte depending on the set source temperature.

ture (TEM). The loss in intensity due to the variation of this parameter is clearly visible here.

We see that the components docetaxel, paclitaxel, and etoposide experience an immense loss in intensity when the temperature is higher than 250 °C. For docetaxel, for example, the intensity is only 10% of its maximum intensity at 450 °C. For almost all other substances, an increase in temperature results in an increase in intensity. As the MS software calculates the optimum temperature on the basis of the sum of the intensities, it suggests a method at 450 °C. This kind of evaluation can therefore lead to certain settings which generate a very high signal for individual analytes and a lower ionization efficiency for all the others. For this reason, the user needs to critically look at the suggested method. If a certain detection limit has to be achieved for all components, a temperature that is too high can lead to a lower signal intensity for temperature-sensitive analytes. However, this would not be visible if the user did not carry out a further data analysis. Furthermore, the user has to choose the settings manually in classical optimization. Because of this, it is possible that the optimum conditions could be not found at all, if the variation of the parameters was set too narrow and the temperature optimization was only carried out for the chromatographic starting conditions as described in Section 3.3.12.2.

In contrast to this, optimization using the QbD approach is based on a few parameters for each variable. Working through the generated experimental design guarantees that the software will ultimately find a global maximum for all components by use of statistical modeling and a search algorithm. Furthermore, it is possible to perform a weighting. Therefore, the optimization of the source parameters can be specifically directed toward the temperature-sensitive components. It is clear from Section 3.3.11.6 that the source temperature for the optimization of the nebulizing gas stream (gas stream 2, GS2) needs to be taken into consideration. Figure 3.30 shows the influence of GS2 on the intensity of the mass transitions for each analyte at two temperature levels.

From Figure 3.30, it is clear that a low GS2 is advantageous for temperature-sensitive substances at a temperature of 450 °C. All other substances exhibit increased intensities with a higher gas stream. When the temperature is reduced to 250 °C, other dependencies appear in relation to GS2. This is presented in Figure 3.30b. A temperature of 250 °C was chosen, as the intensity of the thermosensitive docetaxel is highest at that point. For the substances that are unstable at high temperatures, it becomes evident that the gas flow rate needs to be increased. The maximum intensity is reached for a GS2 between 20 and 30 psi. On the basis of this comparison, it is clear that a change in the source temperature always involves a new optimization of the GS2.

Now we turn to the question of whether the evaluation using a DoE approach delivers similar values for the source parameters in comparison to more extensive manual evaluations. The comparison of values listed in Table 3.5, that optimization using experimental designs enables to determine the optimal range for all substances. The weighted QbD approach, in particular, delivers comparable values for the extensive manual evaluations. In contrast, the method preferred by the

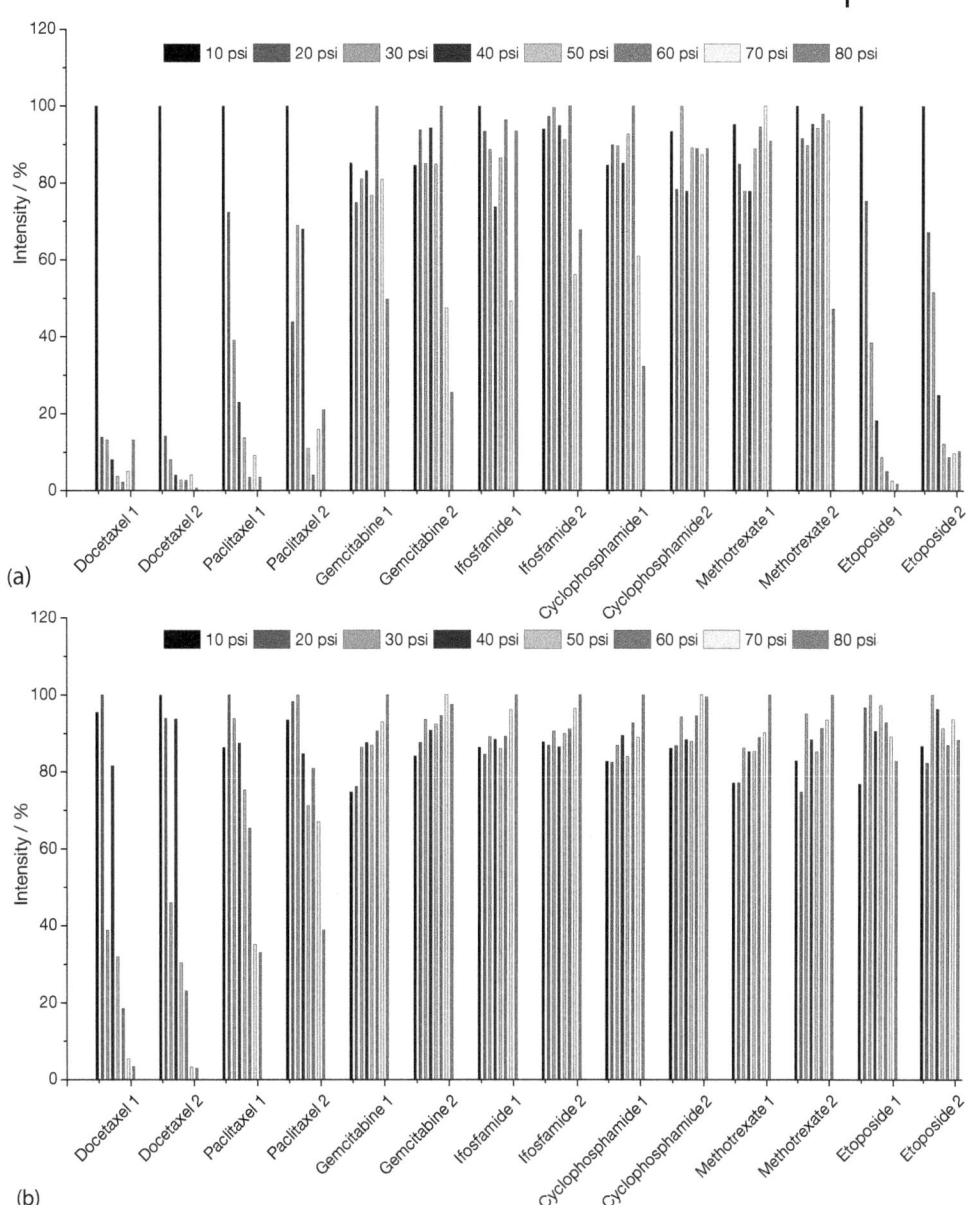

Figure 3.30 Resulting intensities for the two mass transitions of each analyte depending on the setting of the volume stream of the nebulizing gas with source temperature of (a) 450 °C and (b) 250 °C.

Table 3.5 Resulting source parameters depending on the evaluation strategy used.

	MS software	Manual evaluation	QbD approach	Weighted QbD approach
Curtain gas (psi)	20	20	20	20
CAD gas (psi)	High	Medium	Medium	Medium
Ionization voltage (kV)	5.5	5.5	5.5	5.5
Temperature (°C)	450	250	281	263
Gas stream 1 (psi)	20	20	10	10
Gas stream 2 (psi)	60	10	90	28,5

CAD = collision-activated dissociation.

instrument software differs from the optimum parameters quite considerably at this point.

3.3.13
Quantification Using LC/MS

Over the last few years, it has become evident that in the field of target analysis not only triple quadrupole, but also high-resolution time-of-flight mass spectrometers or Orbitraps are being used because the robustness, the linear dynamic range, and the sensitivity have increased immensely. Nevertheless, the number of high-resolution mass spectrometers being used for quantification is still low compared to triple quadrupole devices. Therefore, we would like to discuss the general problems that occur when setting up and applying a multianalyte method for target analysis using these tandem mass spectrometers.

Triple quadrupole mass spectrometers (QqQ) are still the work horses in residue and trace analysis because they allow to record the target analytes selectively, while simultaneously blending out the matrix. As discussed in Section 3.3.3, the compounds contained in the matrix can have a significant influence on the ion suppression. In order to fulfill the requirements of the norm and to ensure reliable quantification results, it is necessary to measure at least two multiple reaction monitoring (MRM) mass transitions per substance, whereby the more intense detected mass transition is used for quantification and the other for substance verification. Acquiring the mass transitions takes time, however, which means that the number of substances that can be recorded in one analysis run is limited. In addition to the acquisition time for a mass transition – called "dwell time" in the literature – it is also necessary to take the "pause time" into consideration. The next mass transition can only be measured after the pause time. By adding all the times together, the so-called "cycle time" is obtained. This is the time needed for generating one data point for each mass transition.

Furthermore, it is often possible to change between the polarities during ionization. In general, one works in the positive ionization mode in ESI; however, some analytes are more sensitive in the negative mode. The question of whether a so-called polarity switching makes sense during an analytical run depends on both the number of compounds to be detected and the time between two peaks, between which the change in polarity is supposed to take place. In addition, the switching time that the mass spectrometer needs for the polarity switching plays a decisive role. In the following example calculations, the change in polarity will not be taken into account as this has an even more unfavorable effect on the number of data points per peak. In contrast, in polarity in methods dealing with only a few target analytes can make a lot of sense, for example, in case a more sensitive or more selective detection is possible in the negative ionization mode than in the positive ionization mode.

The first step in developing an LC/MS multianalyte method is to optimize the mass spectrometric parameters. As noted in Section 3.3.12, the optimization of the parameters that influence the ionization efficiency for a specific mass transition is very complicated. Next, the retention times for all analytes that are to be measured in a multianalyte method should be determined. This can be achieved either by injecting each single standard or the standard mix which contains all components. Once all retention times are known, a calibration can be measured and a linear working range determined. However, this is where the first unpleasant surprises can appear. Figure 3.31 shows the resulting chromatogram of a multian-

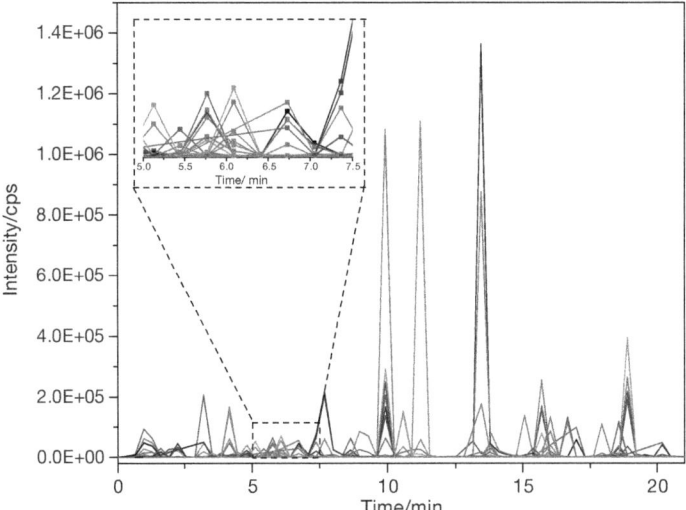

Figure 3.31 LC/MS/MS chromatogram of 182 mass transitions on a 50 × 2 mm Merck Chromolith Fast Gradient RP18 HPLC column. Chromatographic parameters: temperature: 40 °C; injection volume: 20 µL; mobile phases: A = water + 0.1% formic acid, B = methanol + 0.1% formic acid; gradient: 5 to 95% B in 20 min; flow rate: 400 µL min^{-1}; mass spectrometric parameters: pause time: 5 ms; dwell time: 100 ms (unsmoothed raw data).

alyte standard with 91 substances. Looking at the chromatogram, it becomes clear that triangular peak forms occur and only one data point per peak is obtained. Unfavorable settings were obviously chosen for the MS acquisition parameters. So what went wrong here?

In the method presented in Figure 3.31, a total of 182 mass transitions were continuously recorded throughout the whole chromatographic run, where two characteristic mass transitions were chosen for each compound. This so-called retention-time-independent MRM mode has the advantage that it is not necessary to fix the time or time window for the aquisition of the selected mass transitions. Therefore, small fluctuations in the retention times or changes in the solvent gradient do not call for further adjustments in regard to the MS method. However, the disadvantage is that the total number of selected MRM transitions are measured over the complete chromatographic run. On the basis of the following example calculation, it becomes clear that the number of sequentially measured MRM transitions is too high. In the example shown in Figure 3.31, the dwell time was 100 ms and the pause time 5 ms. These settings are the "standard values" as recommended by the manufacturer, for example. They are justified as a long dwell time of 100 ms guarantees that the signal intensity will be increased while the noise will be minimized. If the method only contains a few analytes that are to be quantified, the risk of obtaining an insufficient number of data points per peak is significantly lower. However, in the example selected here, a cycle time of 19.1 s was necessary until all 182 MRM transitions had been measured once, so that one data point could be acquired for each substance. This is equivalent to a data aquisition rate of 0.05 Hz and emphasizes the fact that the default settings need to be critically scrutinized. In principle, it seems that the position where a data point is recorded for a substance is subject to chance. A correct representation of the peak profile, which is demanded for quantification, is not possible. We are currently far from the requirements of the DIN – to generate at least 12 data points across the complete peak. So how do we solve this dilemma?

Hence, attempts are required to adjust the selected parameters for the dwell time and the pause time. If the values are set at 10 ms for the dwell time and 5 ms for the pause time, the resulting cycle time is 2.73 s, which is equivalent to a data aquisition rate of 0.37 Hz. The corresponding chromatogram with the individual data points for all extracted mass transitions is presented in Figure 3.32.

Although we have now managed to increase the number of data points across a peak, we are still far from the requirements specified by the norm. A further reduction in the cycle time is possible with the mass spectrometer being used here but the noise would then increase which means that the method is not suitable for determining very low concentrations. One could consider increasing the peak width for each analyte. Such suggestions were discussed at user seminars when the modern UHPLC systems found their way into routine labs. However, this suggestion would not be advisable as it would not only result in a substantially lower chromatographic separation efficiency and therefore a loss in intensity, but the run time for the method would also increase. This stands in total contrast to the increase in the sample throughput. An alternative is to divide the chromatogram

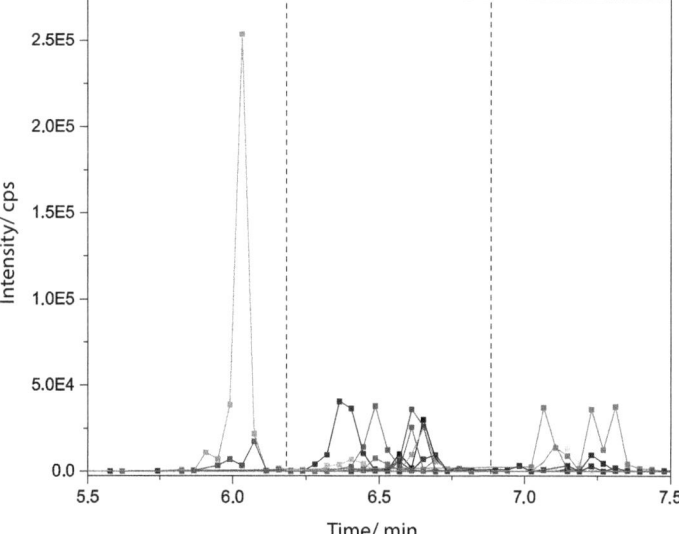

Figure 3.32 LC/MS/MS chromatogram of 182 mass transitions on a 50 × 2 mm Merck Chromolith Fast Gradient RP18 HPLC column. Chromatographic parameters: temperature: 40 °C; injection volume: 20 µL; mobile phases: A = water + 0.1% formic acid, B = methanol + 0.1% formic acid; gradient: 5 to 95% B in 20 min; flow rate: 400 µL min^{-1}; mass spectrometric parameters: pause time: 5 ms, dwell time: 10 ms (unsmoothed raw data).

into periods that are marked by the dashed lines in the chromatogram presented in Figure 3.32. In a specific period, the only MRM transitions that are measured are those of the analytes that elute within that time window. Although by dividing the chromatogram into individual periods, it is possible to reduce the number of MRM transitions that are to be measured sequentially per unit of time, there are a couple of practical pitfalls that need to be considered. To divide the chromatogram into useful periods, time frames need to be identified where no target analytes elute. This is the case in the example presented in Figure 3.32; however, small fluctuations in the retention time can lead to a peak not being recorded within the expected retention time window. Furthermore, as a sample becomes more complex and the number of analytes to be investigated increases, the probability of identifying a region in the chromatogram where no relevant peak elutes sinks considerably. The smaller the chromatographic resolution between neighboring peaks, the more difficult it is to divide the chromatogram into individual periods.

The technical advances made in recent years have led to the use of faster electronics and newer measuring algorithms so that the demand for simultaneous recording of many substances in one chromatographic run has been fulfilled. The solution that MS manufacturers have found for this problem is the development of the so-called retention-time-dependent MRM mode. This is also known as the timed, targeted, or scheduled MRM. Using this MRM mode, it is possible to define an individual aquisition time windows for a specific MRM transition. The great advantage here is that the MRM transitions are no longer measured over the complete chromatographic run, but only within the time frame when the components actually elute from the column. However, typical fluctuations in the retention time for the substance should be considered. For example, small changes in the pH value of the mobile phase have a stronger effect on those substances whose set pH values lie closer to the pK_a value (see explanations in Section 3.3.6.5). It can also be assumed that small changes in the pH value, which can occur due to carelessness while preparing the mobile phase or during a large volume sample injection, lead to marked shifts from the originally determined retention times. During a large volume direct injection, it is the compounds that elute early that are most strongly affected by fluctuations in the retention time. For this reason, one must pay particular attention when determining the detection window for each MRM transition. Sophisticated algorithms do in fact facilitate to define the acquisition time for each MRM transition individually. In this way, there is a drastic reduction in the resulting number of sequentially measured MRM transitions at any one particular time. In contrast to the retention-time-dependent mode, the so-called target scan time needs to be specified by the user. Depending on the number of mass transitions to be measured sequentially, the dwell time is adjusted automatically in such a way that when added to the pause times, the specified target scan time will be achieved. Figure 3.33 shows the chromatogram generated under optimum conditions highlighting the individual data points.

After determining the specific retention time window, one should inspect the region of the chromatogram that exhibits the highest density of peaks which also means the highest number of measured MRM transitions. On the basis of the average peak widths, which can be easily seen on the chromatogram, one can determine the target scan time so that a sufficient number of data points across each peak can be obtained for a successful quantification.

The following calculation is performed to clarify this issue. This is based on two cases. In the first case, it is assumed that a UHPLC system is available and that all critical systems and dead volumes in front of and behind the column are at a minimum. Now we calculate with an average peak width of one second. In the second case, we consider a typical LC/MS coupling, where the distance between the column outlet and the inlet to the ion source is unfavorable. In this case, we calculate with an average peak width of 10 s. A peak width is defined here as the integration points set by the software for the beginning and end of the peak. Wider peaks will not be discussed here as that would lead to an even higher number of data points. We also assume that in spite of using the retention time dependent MRM mode, there are still sections in the chromatogram where 20 substances and, therefore, 40 MRM transitions (!) have to be recorded sequentially. When using older MS devices, the minimum dwell time should not be below 5 ms; with modern devices this can be reduced up to 0.8 ms. However, it is certainly not advisable to use the lowest possible setting, especially in case that the method is designed for determining a low concentration. As a rule, the noise merely increases and the detection limits deteriorate. So, we calculate

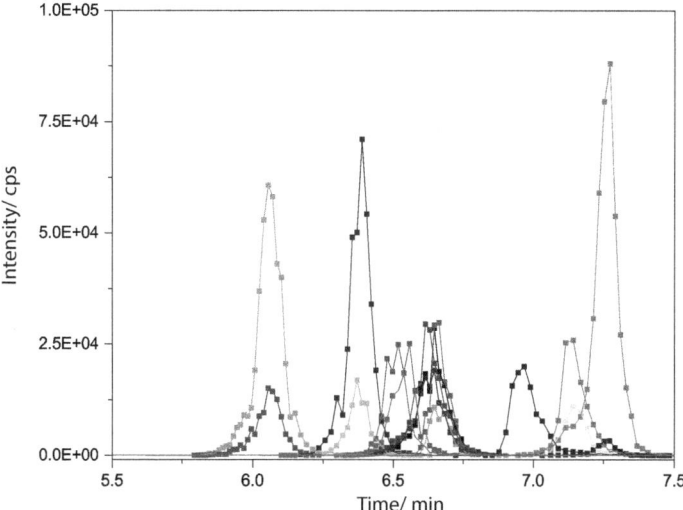

Figure 3.33 Extracted LC/MS/MS chromatogram of 182 mass transitions on a 50 × 2 mm Merck Chromolith Fast Gradient RP18 HPLC column. Chromatographic parameters: temperature: 40 °C; injection volume: 20 µL; mobile phases: A = water + 0.1% formic acid, B = methanol + 0.1% formic acid; gradient: 5 to 95% B in 20 min; flow rate: 400 µL min^{-1}; mass spectrometric parameters: pause time: 5 ms, MRM detection window: 60 s, target scan time: 2 s (unsmoothed raw data).

with a dwell time of 20 ms for older and 2.5 ms for modern systems. Measuring 40 MRM transitions this then leads to an aquisition time of 800 ms and 100 ms respectively. In addition, we have to consider the pause time. Let us also take reasonable values of 5 ms for older and 1 ms for modern devices and so we have total change resulting pause time of 200 ms and 40 ms, respectively. This results in a cycle time of exactly 1 s for older devices and 140 ms for the modern ones, which means a data aquisition rate of 1 Hz and 7.14 Hz, respectively. Using a peak width of 1 s, older mass spectrometers obtain only one data point per peak despite a sophisticated retention-time-dependent MRM mode. In contrast, using modern mass spectrometers results in at least seven data points. Using a peak width of 10 s results in 10 data points for the first scenario and 71 for the second.

As an interim summary, it can be stated that a highly efficient UHPLC separation with very narrow peaks of 1 s in conjunction with older mass spectrometers and a high number of target analytes to be recorded is not an appropriate combination. The peaks generated on the column are simply too narrow. This conclusion does not, however, mean that the combination of a UHPLC system with an older mass spectrometer does not make sense. A big advantage of modern UHPLC systems lies in the consistent reduction of all system volumes. This is why a low gradient dwell volume is an essential requirement to achieve fast analysis cycle times. Particularly, when a relatively low flow rate of, for example, 300 µL min^{-1} is set, it can result in long cycle times with older HPLC systems with a gradient dwell volume of 1 mL, which reduces the sample throughput.

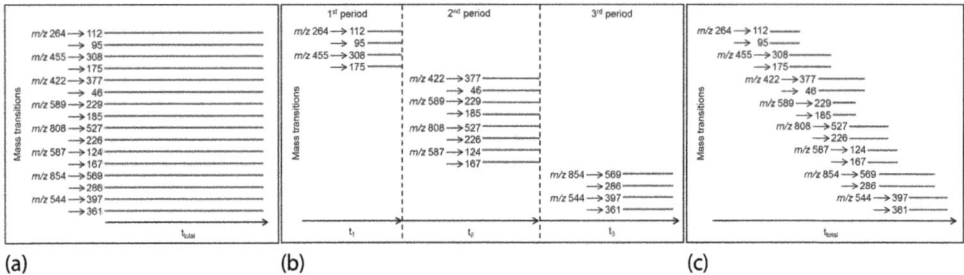

Figure 3.34 Comparative representation of the principle of different MRM modes. (a) Retention-time-independent MRM mode; (b) retention-time-independent MRM mode with the division of the chromatogram in periods; (c) retention-time-independent MRM mode with variable detection windows.

Therefore, taking the specifications of the device and the method into account, the user should determine the number of resulting data points per peak. The average peak width from the chromatogram is to be used as a basis for this. This assumption is justified because, as a rule, the band compression in the solvent gradient mode results in a constant peak width. If, however, peaks elute on an isocratic plateau, for example, at the beginning or at the end of the gradient, an appropriate adjustment to the peak width should be considered for the elution time window. Figure 3.34 shows a comparison of the modes for measuring a specific mass transition explained here in the text.

3.3.14
Screening Using LC/MS

In this section, we take a look at the requirements for suspected-target and non-target screening using LC/MS. The common opinion is that only high resolution systems such as time-of-flight mass spectrometers or Orbitraps can be used for such screening analyses. It must be stated, however, that the characterization of substances with "measuring data information-dependent MS and MS/MS experiments" is also possible using other types of mass spectrometers such as ion traps, triple quadrupole mass spectrometers or "QTRAPs" (quadrupole linear ion trap, QqLIT). These combined measuring algorithms are also known as data-dependent acquisition (DDA), data-independent acquisition (DIA), or information-dependent acquisition (IDA). Recorded data can be evaluated in the same way using all ion fragmentation (AIF). An advantage of this measuring algorithm is that all data can be recorded at all times and no limitation is set, for example, the eight highest intensities or a previously defined positive list, as happens with the data-dependent measuring algorithms. Examples of suspected-target screening are pesticide screening in food control or drug screening in the field of forensic and clinical toxicology [39–42].

In general, all the statements regarding LC/MS coupling that were made in Section 3.3 are also valid for the screening approach. This means that the choice of stationary phase should also be oriented toward the possibilities offered by the mass spectrometer. The same applies for the mobile phase and the gradient run. These statements will not, therefore, be repeated here. An important difference that should be discussed relates to the choice of ionization technology and the appropriate mass spectrometric settings. In screening analyses, there is either no or very little advanced information regarding the expected analytes (see the explanation in Sections 3.2.2 and 3.2.3). Therefore, generic parameters need to be defined in order to record the highest possible number of substances because identification is only possible for the substances that can be sufficiently ionized. This relates primarily to the ionization of the analytes such as the ionization voltage, gas setting, ionization temperature for the evaporation of the solvent, etc.

When screening with HRMS devices, one needs to differentiate between mere MS full scan analyses and combined MS and MS/MS experiments. The maximum sensitivity and mass resolution specified by the manufacturer are generally not achievable at the maximum data aquisition rate. A compromise has to be found here. It makes a significant difference whether the highest possible number of substances based on the molecular formula are to be identified in a chromatographic run using a database comparison or only substances that are present in very low concentration have to be quantified in comparison to the matrix components. With regard to the mass range of m/z 50–1000, relevant for small molecules, on adjusting the mass resolution of the device it is normally possible to record with a sufficiently high scan rate (> 10 data points per peak) using Orbitraps, time-of-flight mass spectrometers, or quadrupole time-of-flight mass spectrometers. According to that, there are no limits with regard to fast chromatography.

However, in many screening approaches the determination of the molecular formula is not a sufficient criterion for a proper identification of substances in complex matrices. Therefore, high-resolution mass spectrometers with the option of generating product ion spectra from precursor ions have been established. Here the rule is: obtaining additional information using, for example, MS/MS spectra needs time. This also means that the manufacturer's specifications regarding the scan rate in mere MS full scan mode needs to be critically questioned or rather corrected. General rules regarding the determination of these parameters cannot be obtained as the "correct" screening parameter is often left to the user's own experience. This abstract behavior needs to be explained in more detail with an example. Figure 3.35 shows the general workflow of a screening analysis.

Figure 3.35a shows the total ion current chromatogram. This results from the sum of all ions recorded in the detector. The total ion chromatogram is of only low significance. In accordance with the set scan rate, there is a complete full scan spectrum for each data point, which is shown as an example in Figure 3.35b. This full scan spectrum shows the recorded mass spectrum across the selected mass to charge range at a specifically chosen time. Due to the m/z ratio and the isotopic pattern as a second criterion beside the deviations in mass, conclusions can be drawn as to the possible molecular formulas of the unknown compounds. However, determining the molecular formula alone is no guarantee that it is in fact the suspected compound. Unlike in GC/MS with its standardized electron im-

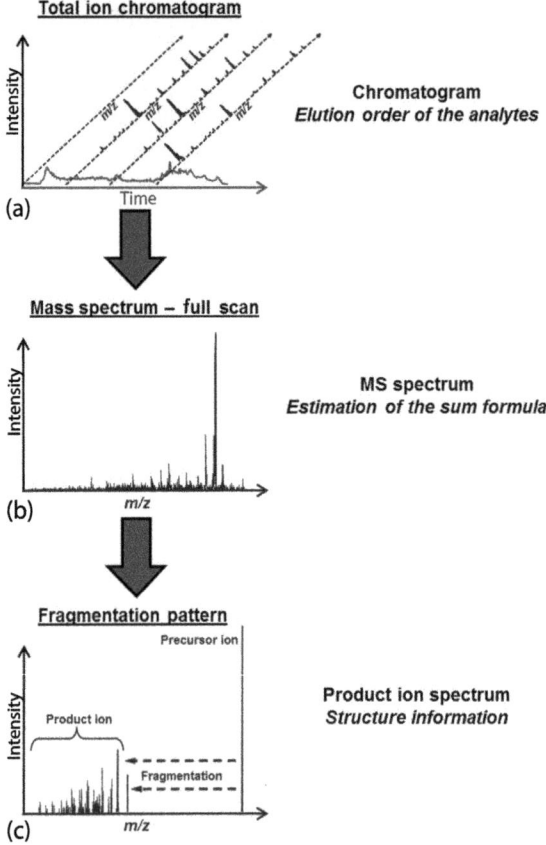

Figure 3.35 General workflow for an LC/MS and MS/MS screening analysis.

pact ionization, the mass spectra generated in LC/MS are heavily dependent on the type of device and the set parameters. In the meantime, it is possible to compare the MS/MS spectra filed in databases from different manufacturers as well as for different types of devices if they were recorded using a standardized protocol (uniform collision energy) [41]. Besides one's own reference database, commercial or free libraries are clever alternatives for comparing data. Freely accessible databases such as CheLIST, Chemspider, DAIOS, Drugbank, HMDB, mzCloud, Norman MassBank, Metfusion, Metlin, PPDB, Stoff-Ident, or TOXNET are gaining in importance. They differ from each other in the number of substances, the number of MS/MS spectra, the meta-information (exact mass, molecular formula, solubility, REACH data, literature references), and the search options [8–14].

In order to further confirm the identity of the suspected compound, one can record product ion spectra. Figure 3.35c shows the schematic representation of this workflow. In information-dependent experiments, for example, the precursor ion with the highest intensity is selected and then fragmented. A product ion

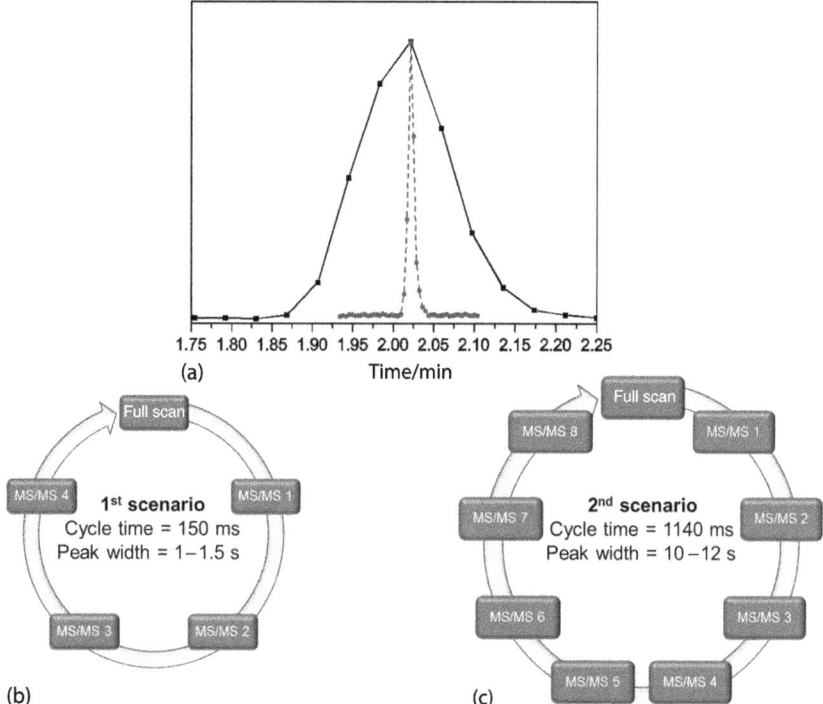

Figure 3.36 Comparison of (a) the obtained peak width using a conventional (black line) and highly efficient (red line) LC separation. In addition, the corresponding cycle times of combined MS full scan and MS/MS acquisition data are shown for (b) four and (c) eight information dependent MS/MS experiments.

spectrum is then obtained that delivers additional information which allows specific conclusions regarding the structure of the compounds.

In combined MS and MS/MS experiments, the number of cycles needs to be defined before a new full scan mass spectrum can be taken. It is useful, for example, to carry out several information-dependent experiments after each full scan rather than just one. Figure 3.36 illustrates the acquisition cycles for four and eight data dependent, information-dependent MS and MS/MS experiments. The number of possible MS/MS experiments is, in turn, dependent on the peak width.

In the first case, besides the full scan, four data-dependent experiments were carried out after the MS full scan measurement. The acquisition time for the full scan was 20 ms, the acquisition time for recording the product ion spectra 20 ms, and the pause time 10 ms. Adding all time values together, we get a cycle time of 150 ms. As can be seen on the basis of the peak profile in Figure 3.36a, about seven to eight data points could be obtained. This design represents all systems that allow highly efficient chromatographic separations in combination with mass spectrometry.

In the second case, on the other hand, eight MS/MS experiments were carried out after a full scan. This was possible because the peaks exhibited a width of about 10–12 s. The acquisition time for the full scan was 250 ms, the acquisition time for recording the product ion spectra 100 ms and the pause time 10 ms. Adding all the time values together, a cycle time of 1140 ms is obtained. The individual peaks could be displayed with about 9 to 10 data points. In contrast to target analysis which has quantification of the substances as the goal, it is not really necessary to display a peak with 12 or more data points in screening analysis. However, if the method is to be used later to facilitate a quantification of known analytes, the MS parameters would have to be adjusted accordingly. Due to the example calculation above, it is clear that for information-dependent MS and MS/MS experiments a suitable increase in acquisition time is necessary so that the number of data points per peak is significantly reduced in contrast to the mere MS mode.

3.3.15
Miniaturization – LC/MS Quo Vadis?

The chapter closes with an outlook toward miniaturization as this topic is gaining in importance even outside academic circles. As regards LC/MS coupling, one can say that instead of 4.6 mm inner diameter columns, columns with diameter of 2.1 mm are becoming more predominant. With regard to the restrictions on the flow rate in ESI, we should mention here that a reduction to an inner diameter of 1.0 mm is desired (see explanations in Section 3.3.9). However, the prevailing opinion is that these columns cannot be packed well enough and therefore exhibit a significantly lower separation efficiency than would be expected theoretically. A central aspect in minimizing the dead volume is the consistent reduction of all volumes of the connecting capillaries. It is often very frustrating that despite all the efforts that an experienced HPLC expert makes, the full use of a highly efficient column is not available in practice. This is always the case when it is not possible to set up the most direct coupling between two modules such as the column oven and the inlet to the ion source (see the system design shown in Figure 3.14). Even when using an optimized system configuration as shown in Figure 3.15, an unfavorable inner diameter of the emitter tip can still lead to noticeable band broadening. This can make the use of columns with an inner diameter of less than 1 mm appear illusionary or visionary.

Therefore, the following experiment should answer the question of whether it is possible to use nano-HPLC columns in combination with a micro-LC system and a conventional mass spectrometer. The stationary phase was a monolithic column from Merck with an inner diameter of 100 μm. This was connected to the injector of the micro-LC system and the emitter tip of the mass spectrometer using a filter. The mass spectrometer being used was an older device from SCIEX (3200 QTRAP). In order to minimize the band broadening after the column, an emitter tip with an inner diameter of 25 μm replaced a conventional tip with an inner diameter of 100 μm. Both variants of tip are shown in Figure 3.37. Whereas the classic tip is made of stainless steel, the modified tip is made from a PEEKSil

Figure 3.37 Comparison of commercially available emitter tips. (a) Classical emitter tip with an inner diameter of 100 µm, compatible with 1/16" fittings; (b) miniaturized emitter tip with an inner diameter of 25 µm, compatible with 1/32" fittings.

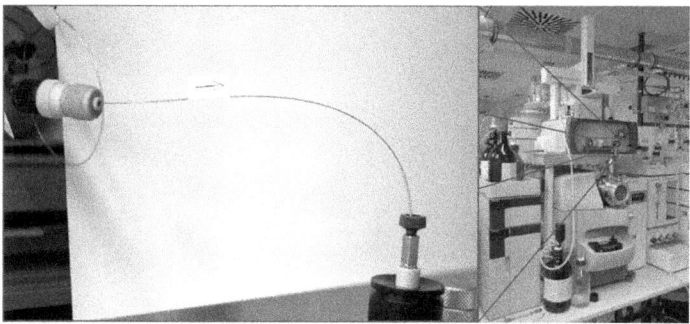

Figure 3.38 Experimental setup for the separation of pharmaceuticals with a monolithic nano-HPLC column (Merck CapROD 150 × 0.1 mm).

capillary. A metal tip of the appropriate inner diameter is installed at the end of the PEEKSil capillary so that ionization is guaranteed. With regard to the connection technology, it is of note that the PEEKSil tips are compatible with 1/32" fittings. High-pressure stable fittings are screwed into a 1/32" connector. This connector (union) also has the advantage that it can be used as a grounding point.

The tip can be changed easily within a few minutes. As can be seen in the system configuration in Figure 3.38, the monolithic column can be used as a transfer capillary between the injector and ion source of the mass spectrometer.

According to the recommendation of the column manufacturer, the pressure should not exceed 200 bar; therefore, under the given conditions, the flow rate was set at 5 µL min^{-1}. Figure 3.39 shows the chromatogram of the separation of approximately 50 pharmaceuticals.

Figure 3.39a shows the separation using a conventional HPLC system and a monolithic phase with an inner diameter of 2.0 mm. In the second case (Figure 3.39b), the system design presented in Figure 3.38 was chosen. When looking at the peak widths, it is noticeable that no significant band broadening occurs on a nano-HPLC column. Initially, this may seem to be a surprising result, especially as no modifications to the ion source of the mass spectrometer are required in order to use it in a low flow range. Furthermore, it needs to be mentioned here that no optimized mass spectrometric conditions were set for measuring with a nano-HPLC column. This is why a "fair" comparison regarding the sensitivity for both approaches (Figure 3.39a). The comparison presented here should simply show the extraordinary potential of the coupling being described. However, it

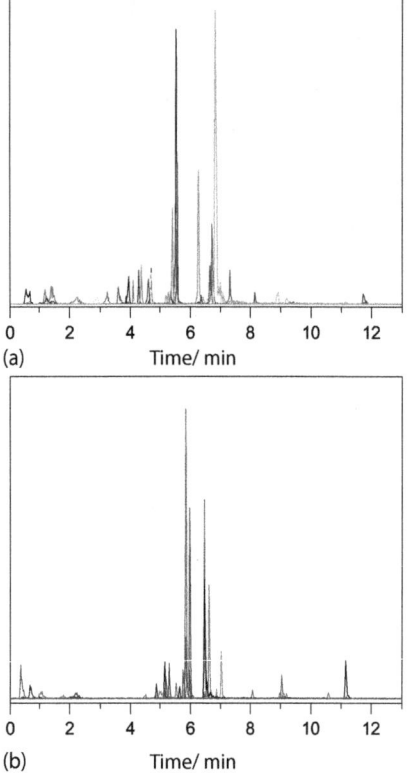

Figure 3.39 LC/MS/MS chromatogram of a separation of 50 pharmaceuticals on (a) a 50 × 2 mm Merck Chromolith Fast Gradient RP18 column; chromatographic parameters: temperature: 40 °C injection volume: 10 µL; mobile phases: A = water + 0.1% formic acid, B = acetonitrile + 0.1% formic acid; flow rate: 500 µL min^{-1}; mass spectrometric parameters: pause time: 5 ms; dwell time: 20 ms; (b) a 150 × 0.1 mm Merck Chromolith CapROD C-18 column; chromatographic parameters: temperature: room temperature; injection volume: 100 nL; mobile phases: A = water + 0.1% formic acid, B = acetonitrile + 0.1% formic acid; flow rate: 5 µL min^{-1}; mass spectrometric parameters: pause time: 5 ms; dwell time: 20 ms.

must be emphasized that the use of the column with an inner diameter of 100 µm is only applicable with systems designed especially for micro- or nano-HPLC, which have been commercially available from numerous providers for several years. In addition, it is absolutely necessary to minimize the dispersion after the column. If the inner diameter of the emitter tip is not adjusted to 25 µm, using nano-HPLC columns with the selected set up is simply not possible.

With regard to miniaturization, it should be noted that conventional unmodified ESI sources work reliably and robustly even with a flow rate of a few µL min^{-1}. Minimizing the flow rate results in a solvent consumption of only 1% if a nano-HPLC column is used. A possibly much more important aspect regarding LC/MS coupling is that less dirt enters the mass spectrometer since the injection volume has to be adjusted accordingly. This is a very important point as separating the matrix from the target analytes presents a big challenge. In practice, it can often be observed that very low detection limits can be achieved for a reference standard that is dissolved in a high purity solvent. Due to the impurities contained in each real sample, more or less distinct matrix effects occur, which lead to ion suppression. This means that miniaturization has an advantage, particularly in the measurement of real samples saddled with a high matrix, as injection volumes have to be reduced. In turn, this also leads to the intervals for cleaning the ion-source becoming significantly longer.

The presentations held at seminars in recent years indicate that all manufacturers are working on the question of miniaturization and that this trend will continue.

References

1 Norm-Entwurf DIN 38407-47 (2015) *Deutsche Einheitsverfahren zur Wasser-, Abwasser- und Schlammuntersuchung – Gemeinsam erfassbare Stoffgruppen (Gruppe F) – Teil 47: Bestimmung ausgewählter Arzneimittelwirkstoffe und weiterer organischer Stoffe in Wasser und Abwasser – Verfahren mittels Hochleistungs-Flüssigkeitschromatographie und massenspektrometrischer Detektion (HPLC-MS/MS oder -HRMS) nach Direktinjektion (F 47)*.

2 International Organization for Standardization (2017) Water quality — Determination of selected active pharmaceutical ingredients, transformation products and other organic substances in water and treated waste water — Method using high performance liquid chromatography and mass spectrometric detection (HPLC-MS/MS or -HRMS) after direct injection, *Committee Draft ISO/CD 21676 from 2017-02-23*.

3 Moschet, C. et al. (2013) Alleviating the reference standard dilemma using a systematic exact mass suspect screening approach with liquid chromatography-high resolution mass spectrometry. *Analytical Chemistry*, **85** (21), 10312–10320.

4 Hug, C. et al. (2014) Identification of novel micropollutants in wastewater by a combination of suspect and non-target screening. *Environmental Pollution*, **184**, 25–32.

5 Martínez Bueno, M.J. et al. (2012) Simultaneous measurement in mass and mass/mass mode for accurate qualitative and quantitative screening analysis of pharmaceuticals in river water. *Journal of Chromatography A*, **1256**, 80–88.

6 Zedda, M. and Zwiener, C. (2012) Is non-target screening of emerging contaminants by LC-HRMS successful? A plea for compound libraries and computer tools. *Analytical and Bioanalytical Chemistry*, **403**(9), 2493–2502.

7 Schymanski, E.L. et al. (2014) Strategies to characterize polar organic contamination in wastewater: Exploring the capability of high resolution mass spectrometry. *Environmental Science and Technology*, **48** (3), 1811–1818.

8 Stoff-Ident database, Bayrisches Landesamt für Umwelt (LfU). Available from: https://www.lfu.bayern.de/stoffident/#!home, accessed on 19/03/2017.

9 Chemspider. Available from: www.chemspider.com/, accessed on 19/03/2017.

10 Daios Online. Available from: www.daios-online.de/, accessed on 19/03/2017.

11 MassBank. Available from: www.massbank.eu/MassBank/, accessed on 19/03/2017.
12 Metlin Scripps. Available from: http://metlin.scripps.edu/index.php, accessed on 19/03/2017.
13 MSBI. Available from: http://msbi.ipb-halle.de/MetFusion/, accessed on 19/03/2017.
14 MzCloud. Available from: www.mzcloud.org/, accessed on 19/03/2017.
15 Human Metabolome Database. Available from: www.hmdb.ca/, accessed on 19/03/2017.
16 Pesticide Properties Database. Available from: http://sitem.herts.ac.uk/aeru/ppdb/en/index.htm, accessed on 19/03/2017.
17 Drugbank. Available from: www.drugbank.ca/, accessed on 19/03/2017.
18 Toxicology Data Network. Available from: http://toxnet.nlm.nih.gov/, accessed on 19/03/2017.
19 Chemical Lists Information System. Available from: http://chelist.jrc.ec.europa.eu/, accessed on 19/03/2017.
20 Wolf, S. et al. (2010) In silico fragmentation for coumputer assisted identification of metabolite mass spectra. *BMC Bioinformatics*, **11**, 148. .
21 Davis, J.M. and Giddings, J.C. (1983) Statistical theory of component overlap in multicomponent chromatograms. *Analytical Chemistry*, **55** (3), 418–424.
22 Dolan, J.W. (2007) LCGC Europe, **21**, 386.
23 Teutenberg, T. (2010) *High Temperature Liquid Chromatography – A User's Guide for Method Development*, Royal Society of Chemistry, Cambridge.
24 Teutenberg, T. et al. (2009) High-temperature liquid chromatography. Part II: Determination of the viscosities of binary solvent mixtures – Implications for liquid chromatographic separations. *Journal of Chromatography A*, **1216** (48), 8470–8479.
25 Hetzel, T., Teutenberg, T., and Schmidt, T.C. (2015) Selectivity screening and subsequent data evaluation strategies in liquid chromatography: the example of 12 antineoplastic drugs. *Analytical and Bioanalytical Chemistry*, **407** (28), 8475–8485.
26 Kowal, S., Balsaa, P., Werres, F., and Schmidt, T.C. (2012) *Analytical Bioanalytical Chemistry*, **403** (6), 1707–1717.
27 Neue, U.D. and Alberto, M. (2007) Selectivity in reversed-phase separations: General influence of solvent type and mobile phase pH. *Journal of Separation Science*, **30**, 949.
28 Neue, U.D. (2008) Peak capacity in unidimensional chromatography. *Journal of Chromatography A*, **1184** (1–2), 107–130.
29 Subirats, X., Rosés, M., and Bosch, E. (2007) On the effect of organic solvent composition on the pH of buffered HPLC mobile phases and the pK a of analytes – A review. *Separation and Purification Reviews*, **36** (3), 231–255.
30 Liverpool, U.O. (2006) Available from: https://www.liverpool.ac.uk/buffers/buffercalc.html.
31 Beynon, R.E.J. (2003) *Buffer Solutions – The Basics*, Taylor & Francis Group.
32 Kromidas, S. (2006) *HPLC richtig optimiert*, Wiley-VCH Verlag GmbH, Weinheim.
33 Fekete, S., Oláh, E., and Fekete, J. (2012) Fast liquid chromatography: The domination of core–shell and very fine particles. *Journal of Chromatography A*, **1228**, 57–71.
34 Hayes, R. et al. (2014) Core–shell particles: Preparation, fundamentals and applications in high performance liquid chromatography. *Journal of Chromatography A*, **1357**, 36–52.
35 Leonhardt, J. et al. (2014) Large volume injection of aqueous samples in nano liquid chromatography using serially coupled columns. *Chromatographia*, **78** (1), 31–38.
36 Li, Y., Whitaker, J.S., and McCarty, C.L. (2012) Analysis of iodinated haloacetic acids in drinking water by reversed-phase liquid chromatography/electrospray ionization/tandem mass spectrometry with large volume direct aqueous injection. *Journal of Chromatography A*, **1245**, 75–82.
37 Snyder, L.R. and Dolan, J.W. (2007) *High-Performance Gradient Elution – The Practical Application of the*

Linear-Solvent-Strength Model, Wiley-Interscience, John Wiley & Sons.

38 West, C., Elfakir, C., and Lafosse, M. (2010) Porous graphitic carbon: A versatile stationary phase for liquid chromatography. *Journal of Chromatography A*, **1217** (19), 3201–3216.

39 Alder, L. *et al.* (2006) Residue analysis of 500 high priority pesticides: Better by GC-MS or LC-MS/MS? *Mass Spectrometry Reviews*, **25** (6), 838–865.

40 Dresen, S. *et al.* (2009) ESI-MS/MS library of 1253 compounds for application in forensic and clinical toxicology. *Analytical and Bioanalytical Chemistry*, **395** (8), 2521–2526.

41 Oberacher, H. *et al.* (2012) On the inter-instrument and the inter-laboratory transferability of a tandem mass spectral reference library. 3. Focus on ion trap and upfront CID. *Journal of Mass Spectrometry*, **47** (2), 263–270.

42 Oberacher, H., Weinmann, W., and Dresen, S. (2011) Quality evaluation of tandem mass spectral libraries. *Analytical and Bioanalytical Chemistry*, **400** (8), 2641–2648.

Part II
Tips, Examples, Trends

4
LC/MS for Everybody/for Everything? – LC/MS Tips
F. Mandel

4.1
Introduction

If you are already using a LC/MS system or you are planning to do so, you will have good reasons for it, but it does not mean you should discard "traditional" optical detection techniques altogether. While detection limit, selectivity, and strength of evidence of LC/MS(/MS) may be unequaled, the initial investment in equipment and time are considerable before a rock-solid LC/MS method has been established. In general, all LC/MS detection methods are less linear than UV detection and they provide a smaller dynamic range and lower reproducibility. If you determine compounds in complex sample matrices, you should spend some thoughts on interferences of the sample matrix and the analyte signal.

Why this substantial investment in manpower and machinery? It will save you time when developing new detection methods. This is no contradiction to what has been said above because while it make take longer to develop a detection method, this extra investment is compensated many times over by the resulting simplification of sample preparations and chromatographic separations. LC/MS is the method of choice when it comes to quick chromatography of complex samples. Those of us who spent most of their career perfecting the art of baseline separation will find that LC/MS means a change of paradigm – the quantification of co-eluting peaks. This means that the buffer system has to be selected and optimized in view of the detector rather than chromatographic separation. Compared to UV detection, LC/MS offers not only more rapid separations but also higher detection sensitivity. Fragmentation patterns render unambiguous identification of substances possible. However, the LC/MS spectral libraries are still in their early stages – what is lacking is a standardization of instrumental parameters.

Similar to driving a car, where you won't reach your destination without having considered some rules of physics, you cannot "just start" with LC/MS. In the following LC/MS tips, we will show you how to find the LC/MS technique that fits best your analytical problem. You will also learn about the strengths and weaknesses of the different LC/MS techniques as well as to recognize possible sources of error and how to minimize them. As you may know already – the LC/MS tech-

---

The HPLC-MS Handbook for Practitioners, First Edition. Edited by S. Kromidas.
© 2017 WILEY-VCH Verlag GmbH & Co. KGaA. Published 2017 by WILEY-VCH Verlag GmbH & Co. KGaA.

nique that fits everything does not exist, unfortunately. You have the opportunity to choose from various ionization techniques and mass analyzers, the application ranges of which mostly overlap. This is why we will compare the most common ionization techniques and discuss their advantages and disadvantages in the following sections. No need to be scared, this will not end up in a lesson of ion physics – it will be a rather pragmatic approach to questions like "how does the charge get on my molecule" and "how can I detect the charged molecule". My intention is to describe not only "how it works" but also "why", in order to enable you to develop and optimize your methods on your own, without having to consult books or ask for assistance most of the time.

LC/MS is no longer the exotic technology it once was, and some of the mass spectrometers available nowadays hardly take up more bench space than an HPLC system. Although it began in the 1970s, it was only in the early 1990s that instruments and ionization techniques suitable for routine work became widely available. The big breakthrough happened with the use of atmospheric pressure ionization techniques, which earned John Fenn the Nobel Prize in chemistry in 2002. LC/MS can be used to quantify trace amounts, collect fractions based on mass, elucidate molecular structures by MS^n, identify even the smallest amounts of proteins by using nano-HPLC techniques, and much more. Nowadays, the user interfaces of the data systems are very easy to operate, including "open access" or "walk-up" mass spectrometry, and for many users, the LC/MS system seems to have become a "black box". In the following tips, I invite you to have a look into the inner workings of your LC/MS system.

4.2
Tip Number 1

4.2.1
Choosing the Right LC/MS Interface

In all decisions you take in LC/MS, the most important criterion is ionization. Whatever you may undertake in order to achieve highly sensitive, highly selective, and/or high-resolution measurements – it is most important to ionize the analytes as gently as possible. A mass spectrometer is only capable of measuring either positively or negatively charged molecules ("pseudomolecular ions"). Unfortunately, very often, the chemistry of these ions does not make it easy for you – the analytes to be measured can cover a wide range of polarity and molecular weight. They may also be more or less thermolabile (otherwise we would use GC/MS instead of LC/MS). In addition, a mass spectrometer has to be operated at high vacuum. Therefore, we have to evaporate the mobile phase first and then remove it from the analytes ("desolvation"). As we will see in the following sections, some LC/MS interfaces are capable of handling high LC flow rates, while others show their strengths at low flow rates. Whatever technique you choose, it will be an API technique (atmospheric pressure ionization) like electrospray (ESI),

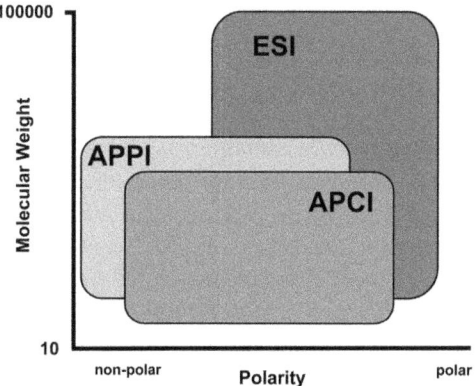

Figure 4.1 Application ranges of API techniques.

atmospheric pressure chemical ionization (APCI), or atmospheric pressure photoionization (APPI) (Figure 4.1). All other interfacing techniques (thermospray, particle beam) are either of just historical interest or they are not suitable for routine work.

4.2.1.1 Polar, Thermolabile Analyte, High Molecular Weight – Shall I Use Capillary or Nano-HPLC?

Try electrospray first! Before you start, you should understand how it works (Figure 4.2). Actually electrospray is not a method for ionizing analytes, but a way of releasing cations or anions that have been formed in the HPLC buffer system by the acid/base equilibrium. This aspect is of fundamental importance when you

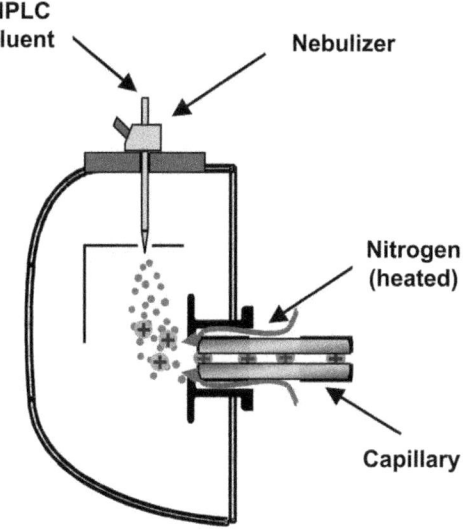

Figure 4.2 Schematic representation of an electrospray ion source.

develop your LC/MS methods or when you troubleshoot your system. You should therefore choose pH conditions that force the formation of free analyte cations or anions in solution. The selection of "MS compatible" mobile phases will be the topic of one of the tips to follow.

Electrospray usually utilizes a pneumatic nebulizer in order to convert the eluent into an aerosol. This is also called "pneumatically assisted electrospray" or "IonSpray®". The aerosol is then sprayed into an electrostatic field. A high voltage of several kilovolts is applied between the nebulizer and the entrance of the ions into the vacuum system ("orifice"). Depending on the polarity of the charged analyte molecules, the resulting aerosol droplets carry a positive or negative net charge. This causes them to be attracted toward the orifice or the ion transfer capillary where they are transferred to the ion optics of the mass spectrometer. It is important not to introduce the charged droplets into the mass spectrometer. This would trigger a high background signal and therefore negatively affect detection sensitivity. Modern ion source designs are therefore based on an orthogonal geometry, spraying the eluent at a 90° angle onto the ion entrance and the ion path. Simultaneously heat is applied via a stream of heated nitrogen, in order to evaporate the solvent from the aerosol droplets. On their way toward the orifice or ion transfer capillary, the droplets undergo a cascade of shrinking and explosions into smaller droplets ("Coulomb explosions") in order to stabilize the droplet's surface charge. In addition, (pseudo)molecular ions are spontaneously emitted from the droplets. Orthogonally designed ion sources ensure that almost exclusively desolvated (pseudo)molecular ions find their way into the mass spectrometer. Because of their high mass, residual droplets cannot be deflected by 90°. Orthogonal ion sources do not only exhibit a high level of sensitivity, but also of robustness and are only marginally affected by contamination.

Which flow rates can be handled by electrospray ion sources? Pneumatically assisted sprayers operate in a range of 0.5 µL/min to 2 mL/min. Of course this could involve a variety of geometries and/or nebulizers. For flow rates of 0.5 µL/min to approximately 50 µL/min, sprayers are used that have been optimized for extremely low dead volumes ("microspray"), while a standard sprayer is used for higher flow rates. Depending of the manufacturer, these provide optimum ion yields within a flow rate range of 100 µL/min to 1 mL/min. We have learned already – in electrospray the ions are formed in solution. Therefore, it is a concentration-dependent detection technique, like UV detection. This has the advantage that the eluent stream can be split in front of the mass spectrometer without losing sensitivity. This is very useful when using HPLC columns with large diameters and the high HPLC flow rates associated with them, when using two detectors simultaneously or when MS-based fraction collection of the eluent stream is required ("mass-directed" or "mass-based" fractionation).

Smallest sample amounts and volumes, as in "proteomics" applications, require a highly sensitive way of detection. Currently, the most sensitive LC/MS technique is using a nanospray in conjunction with nano-HPLC. Nanospray is actually a variation of the original "classical" electrospray, where the aerosol is formed solely by high electrical fields, without any pneumatic support. At the moment, the

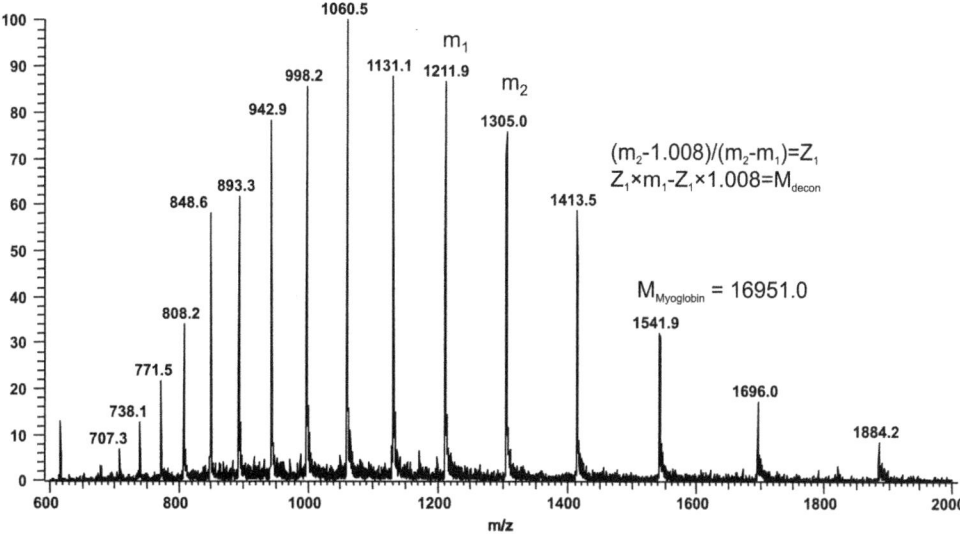

Figure 4.3 Electrospray mass spectrum of myoglobin.

use of nanospray is restricted to the detection of biopolymers (peptides, proteins, oligonucleotides). With increasing robustness it could be transferred to other application areas as well.

An important characteristic of electrospray is the formation of multiply charged (pseudo)molecular ions – depending on the number of basic (positive ESI) or acidic (negative ESI) functional groups and the HPLC buffer system. Why is this so important? Any mass spectrometer has a limited (physical) mass range. The upper limit usually is around m/z 4000 to m/z 6000. The exceptions are time-of-flight mass spectrometers with high acceleration voltages. You probably know, in mass spectrometry, it is always the mass/charge ratio that is determined. Only the ability of electrospray to form multiply charged ions enables us to measure proteins of molecular weights above 100 000 Da. Figure 4.3 shows the mass spectrum of a small protein. Each peak represents another charge state of the molecular ion.

4.2.1.2 Low Polarity, High HPLC Flow Rates, Volatile Analyte?

In that case, APCI is your interface of choice. The difference in construction between APCI and electrospray is minimal, however essential for the ionization process. Unlike in electrospray, we do not make use of the ions formed in solution, but generate them in the gas phase instead. The aerosol formed from the mobile phase and the analyte is sprayed into a heated ceramic or glass cartridge and completely evaporated. That vapor is sent past a needle, to which several kilovolts are applied, and at the tip of which a corona discharge is formed. The plasma produced in the process enables chemical ionization to take place. While colliding with the (still) neutral analyte molecules, the ionized solvent molecules act as pro-

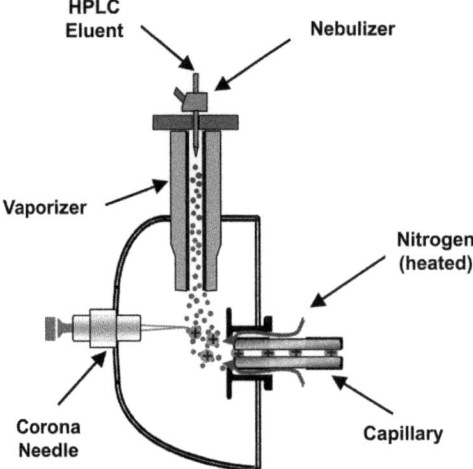

Figure 4.4 Schematics of an APCI ion source.

ton donors or acceptors. The difference in proton affinities between the molecules colliding with each other is important. With electronegative analyte molecules we can also observe electron capture processes. As the charges are transferred to the analyte molecule through single collisions, we generate only singly charged molecular ions in APCI. Of course, this limits the detectable molecular weight range compared to electrospray. A further restriction in the application of APCI is the need to evaporate the analyte without thermal degradation – a step many thermolabile compounds do not survive. Therefore, the vaporizer temperature is an important parameter in APCI method development (Figure 4.4).

What are the reasons in favor of APCI as a detection technique in LC/MS? It is ideally suited for high HPLC flow rates. Most APCI source designs reach their optimum performance at 0.5 to 1.5 mL/min. A great advantage of APCI is its ability to ionize analytes of weak polarity, which are not accessible in electrospray at all or only under extreme pH conditions. In APCI the decisive factor is not pH value of the mobile phase, but the gas phase acidity/basicity. This allows you to optimize the pH of the mobile phase solely for chromatographic separation without having to consider the mass spectrometer as well. In addition, in APCI you will not find "mixed ionization" caused by protonation and alkali adduct formation. As we will see in one of the subsequent tips, APCI methods show a more linear response than Electrospray methods and also suffer less from ion suppression effects.

4.2.1.3 Still No Signal, Analyte Difficult to Evaporate or Nonpolar, APCI Not Sensitive Enough?

Although it is not yet a widespread method, APPI could be the answer to your problems in analysis. Atmospheric pressure photoionization is a slight modification of APCI. After the evaporation step, the analyte and the mobile phase sent past a krypton UV lamp emitting light energy of 10 and 10.6 eV. This excitation

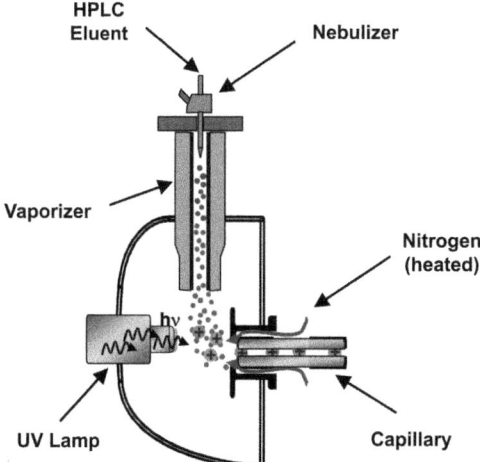

Figure 4.5 Schematics of an APPI ion source.

energy induces photoionization of the analytes, the ionization energy of which, of course, has to be below 10 or 10.6 eV, respectively. The widely used HPLC mobile phases, such as water, methanol, acetonitrile and hexane, exhibit ionization energies above the excitation energy and will thus not be ionized. This first photoionization step leads to the formation of a radical cation, to which a hydrogen atom originating from a nonionized solvent molecule may be transferred. Sometimes, the radical cation and the molecular cation can be detected alongside each other. Should the detection sensitivity of that direct photoionization not be sufficient, you can add easily ionizable modifier ("dopants") to the mobile phase directly in front of the ion source (about 5% v/v to the eluent). The photoionized dopant then serves as a proton donor, ionizing the analyte. Typical dopants or modifiers are acetone and toluene. Acetone is also an excellent donor of thermalized electrons, which can be used to form analyte anions after an electron capture step. The optimum HPLC flow rates for APPI are in the range of 0.5 mL/min (Figure 4.5).

It is amazing what nonpolar compounds can be analyzed using APPI. It is the one and only ionization technique that enables the sensitive cationization of polycyclic aromatic hydrocarbons. In contrast to APCI, the ionization process in APPI is not negatively affected by residual aerosol droplets. Because of this, the vaporizer can be operated at lower temperatures. Of course, this helps in the detection of thermolabile compounds that are not detectable in electrospray. In general, the sensitivity of APPI is quite comparable to APCI. However, in APPI the range of analyte polarity and thermolability is significantly wider than in APCI.

4.3
Tip Number 2

4.3.1
Which Mobile Phases Are Compatible with LC/MS?

You may have anticipated it already – as an LC/MS user you will have to say goodbye to a good old friend – the phosphate buffer. As you will see in a minute, there are exceptions to any rule. However, when converting existing HPLC methods to LC/MS or by developing new methods, you should base all of them on volatile buffer systems. Before we start discussing buffers – let us talk about which solvents work best with which LC/MS interface.

4.3.1.1 The Solvents

The following solvents are compatible with electrospray and APCI: alcohols, acetonitrile, tetrahydrofuran (THF), water, acetone, dimethylformamide, methylene chloride, and chloroform. If, in APCI, you at least partially replace acetonitrile by methanol, you will enhance detection sensitivity as well as the long-term stability of the analyte signal. This is because gaseous acetonitrile is a relatively strong base and therefore competes with the analytes for protonation. In addition, acetonitrile tends to polymerize in the APCI plasma, coating the corona needle with an insulating layer after some hours of operation. Consequently, more frequent abrasive cleaning of the APCI needle will be required. Dimethylformamide should be below 10% v/v when running API electrospray, while in APCI you should be prepared for a high background signal. In APCI, THF also tends to polymerize, in particular when it contains traces of peroxides. There is a trick to minimize the precipitate in the APCI ion source. Right before the mobile phase enters the ion source, you should add about 5% v/v of water. This also stabilizes the corona discharge while running THF. Halogenated hydrocarbons can enhance the ion yield when used as modifiers in APCI. In electrospray, they show neither a positive nor a negative influence. In general, the less protic your solvent is, the less it will be suited for electrospray (acid/base equilibrium).

In APCI, we make use of the mobile phase as a "reagent gas" for chemical ionization of the analyte. Therefore, besides the ones already mentioned, aliphatic and aromatic hydrocarbons are also permitted, as well as CS_2 and CCl_4. Toluene is an excellent proton donor in APCI.

4.3.1.2 The Additives

What should be used instead of phosphate buffers? The most important rule is to use volatile buffer additives and to use organic acids. I can hear the screams of protest of the experienced HPLC user. But, to be honest, most of the analytes can be separated successfully in an "MS-compatible" way by using the modern RP column materials. Please take into account that mass spectrometers are able to differentiate between compounds by the m/z signal if they cannot be separated chromatographically. For your validated methods which use phosphate buffers, I will tell you what the exception from that rule is.

But let us discuss the additives first: use ammonium acetate or formate to buffer the pH. Acidify the pH by adding acetic acid, formic acid, trifluoroacetic acid (TFA) when running positive ion electrospray and adjust a basic pH in negative ion electrospray by adding ammonia, triethylamine (TEA) or N-methylmorpholine. Another rule is to use a buffer concentration as low as possible – below 10 mmol/l in electrospray and maximum 100 mmol/l in APCI or APPI. Please take into account that TFA is a weak ion pairing reagent and, therefore, reduces the detection sensitivity for many analytes. Please avoid the use of even low TFA concentrations if you frequently change ion polarity in electrospray. The TFA anion gives a permanent background signal at m/z 113. If you need a basic pH for either the chromatographic separation and/or the ionization, you should use ammonia instead of TEA. Ammonia does not show any memory effect in your HPLC system while TEA shows a background signal at m/z 102 during subsequent measurements in positive ion electrospray. Both with TFA and TEA, it could take days until the background signal has dropped to an acceptable intensity level.

Why all that worry about volatile buffers and low buffer concentrations? After evaporation of the solvents, nonvolatile buffers, will precipitate inside your ion source and, depending on the ion source geometry, will soon block the ion entrance or will cause leaking current or shortages. Even if this does not occur, the alkali cations will block the aerosol droplet surfaces (potassium or sodium phosphate) in electrospray ionization and hinder the emission of the analyte ions. The need for low buffer concentrations is easily explained as well. With all API techniques a cloud of charged species is formed together with ionized sample matrix and buffer additives. The charge density in that cloud is limited by space charging (distraction) of species of the same polarity. Finally, this leads to a widened spray and a "dilution" of the analyte ions in the cloud at higher buffer concentrations. As the spray is more collimated in electrospray than in APCI or APPI, the latter two are more suited for high buffer concentrations.

4.4
Tip Number 3

4.4.1
Phosphate Buffer – The Exception

Please do not interpret the following hint as a general permit to follow old habits of working with phosphate buffers. Do not deviate from the iron rule to develop any new LC/MS methods by using volatile buffers. But what if you have to investigate a chromatographic peak – and unfortunately the method had been extensively developed and/or being validated using phosphate buffers? In this case, you might be allowed to work with the "forbidden" buffers. However, you should be prepared for a heavy contamination of the spray chamber and a major cleaning-up effort.

Many ion source designs will allow you to keep your tools at hand. Please consult also the manufacturer of your mass spectrometer or other users of the same instrument type.

4.4.1.1 How Does It Work?

Please remember the first tips. If you want to determine analyte cations, then electrospray suffers from a dramatic loss in detection sensitivity with sodium or potassium phosphate buffers. The reason for this is the suppression effect caused by the alkali cations in the aerosol. Therefore you should select APCI as an ionization technique while using these nonvolatile buffers. In APCI, both analyte and mobile phase are evaporated before the ionization step. Of course, nonvolatile buffer components will precipitate as a white powder in your ion source. Nevertheless, the analytes will be protonated in the corona discharge. Because of the severe contamination of the ion source, your LC/MS system will "survive" that procedure just a few hours or even only for a single chromatographic run, depending on the buffer concentration. When looking for analyte anions however, you should choose electrospray instead of APCI. The reason is that the phosphate anion is volatile enough to be eliminated from the aerosol droplets similarly to the analyte anions. But please be aware of some reduction in sensitivity. The mass spectrometer will "forget" that phosphate buffer treatment rapidly. Your HPLC system and your column, however, will deliver alkali cations even weeks later, causing unwanted and very often disturbing adduct formation.

4.4.1.2 Summary

In few exceptional cases (!) you can operate your LC/MS system with nonvolatile buffers. Use APCI for the detection of cations and electrospray for the detection of anions. Be prepared for a rapidly increasing contamination of your ion source as well as for the frequently needed cleaning procedures. Preferably use a separate HPLC system with nonvolatile additives.

4.5 Tip Number 4

4.5.1 Paired Ions

Besides phosphate buffers there are other troublemakers of comparable impact – the ion pair reagents. Like nonvolatile buffer systems, ion pair reagents significantly reduce detection sensitivity. Do not get confused by publications stating the opposite – if analytes do not form ion pairs, naturally, their ionization will not be negatively affected. But why do you want to use this kind of buffer additives? You might want to make highly polar compounds more lipophilic rather than ionic in the mobile phase. The effect is an increased retention on reverse phase columns and less peak tailing. However, as you know already, electrospray ionization is based on the release of ions from the aerosol droplets. When masking the analyte ion by a counter ion you will not be able to detect your analyte, unless you break the ion pair apart through the application of heat.

4.5.2
Which "Antidote" Is Available?

You should avoid strong and nonvolatile ion pair reagents, such as tetrabutylammonium bromide or heptanesulfonic acid. Not only does their usage drastically reduce sensitivity – being themselves ions in solutions – they contaminate every HPLC system for a long time and cause a high background signal. For example, even ultralow traces of the tetrabutylammonium cation result in an intense signal at m/z 242. Even the weak ion pair reagent TFA, which is widely used in peptide analysis, will significantly suppress the signal of basic compounds such as the LSD (lysergic acid diethyl amide) metabolite LAMPA (lysergic acid methylpropyl amide), LSD itself is not negatively influenced by TFA.

You can counteract the suppression effect of TFA by adding a high concentration of organic acid (i.e., 50% v/v of propionic acid in isopropanol) right in front of the electrospray ion source, in order to displace the TFA anion from the ion pair. If you cannot avoid the use of ion pair reagents, you should take their volatility into account. Acidic functional groups can be properly masked by aliphatic amines, like triethylammmonium acetate, n-butyl-dimethylammonium acetate or di-n-butylammonium acetate. On the other hand you can replace alkanesulfonic acids with perfluorinated organic acids of different aliphatic chain lengths. In most cases, even in electrospray, there will be just very little suppression of detection sensitivity. If possible you should opt for APCI or APPI, as the weak ion pairs just described will be cleaved by the evaporation process. Please be prepared for a significant background signal when you invert the ion polarity after your ion pairing experiments. As when using high alkali concentrations – that is, phosphate buffers – you should mark the bottles and HPLC columns used in connection

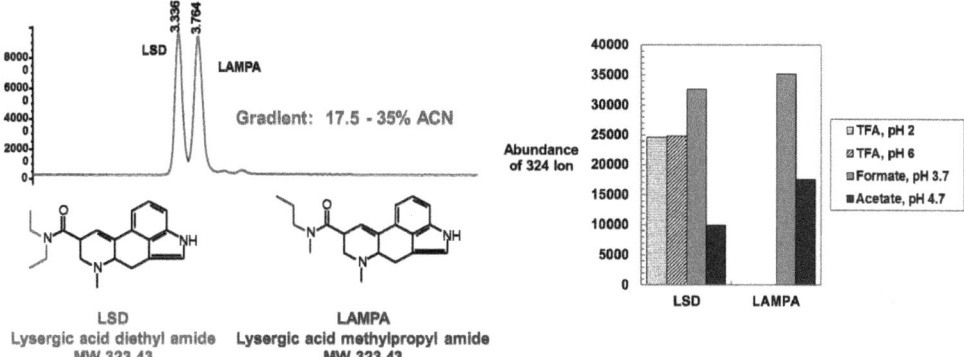

Figure 4.6 Influence of trifluoroacetic acid (TFA) on LSD (lysergic acid diethyl amide) and LAMPA (lysergic acid methylpropyl amide).

with the ion pair reagents and never ever use them with "normal" mobile phases. Moreover, it is not a good idea to put these "contaminated" mobile phases through a vacuum degasser, which has typically a large inner surface consisting of porous Teflon and would thus produce an unpleasant memory effect.

4.5.3
Summary

Traditional ion pair reagents are not suited for use in LC/MS. It is better to use volatile reagents like aliphatic amines or perfluorinated organic acids at the lowest possible concentration. Avoid contamination of your vacuum degasser and use solvent bottles and HPLC columns that are dedicated to the use with ion pair chromatography.

4.6
Tip Number 5

4.6.1
Using Additives to Enhance Electrospray Ionization

In LC/MS as in everything else in life, it is the dosage that matters. You can enhance ionization and detection sensitivity by adding Na or K ions to the eluent, as long as you obtain the correct concentration inside the ESI ion source – it should be in the range of 0.5 mmol/l (Figure 4.7). Please make use of postcolumn addition and do not add the modifier to your mobile phase upfront. This will avoid bad surprises by contamination of your HPLC system.

Cationization through alkali is ideally suited for all compounds that consist of several OH functional groups, such as carbohydrates or steroids. These are difficult to protonate and, thus, not very sensitive to electrospray detection. By adding modifiers you obtain a uniform ion formation and, therefore, a higher sensitivity and increased reproducibility. Figure 4.7a shows that prednisolone produces mul-

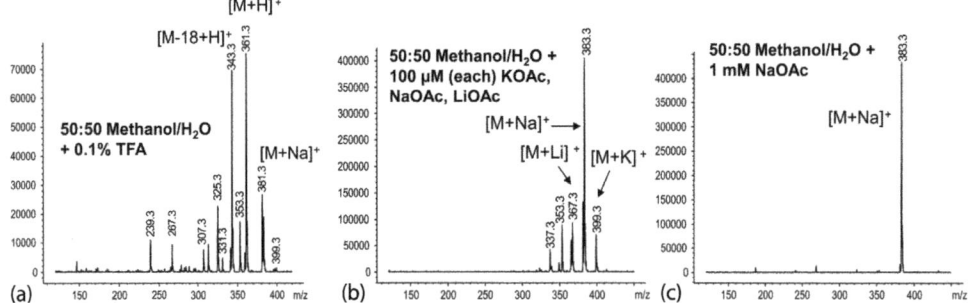

Figure 4.7 Adduct formation in ESI.

tiple molecular cations, even when acidified with TFA. The intensity is distributed across many molecular cations, the ratio of which can vary significantly depending on the mobile phase composition. Therefore reproducible quantitation is almost impossible. In Figure 4.7b, an equimolar mixture of different modifiers was added in order to evaluate their affinities to prednisolone. It can be easily seen that the Na adduct provides the most intense signal. Figure 7c shows the result of postcolumn addition of Na acetate. Now we have a uniform ionization and a stable pseudomolecular ion. But not only alkali salts make carbohydrate detection easier in ESI, you can also add 50 mM HCl postcolumn (concentration in the ESI source about 2 mM/l) and force the formation of chloride adducts. These are of course detected in negative ESI mode.

4.6.2
Additives for APCI

Let us stick to the prednisolone example. Normally it gives a $[M + H - H_2O]^+$ signal in positive ion APCI, while not being detected in negative ion APCI. However, if you add 1–5% v/v methylene chloride postcolumn, you obtain a very intense $[M + Cl]^-$ ion (Figure 4.8). Phenolic compounds can be detected with enhanced sensitivity in negative ion APCI as soon as traces of either oxygen or trichloromethane are present. A very common APCI modifier is toluene, which, as a postcolumn addition to the eluent in approximately 5% v/v, makes an excellent proton donor.

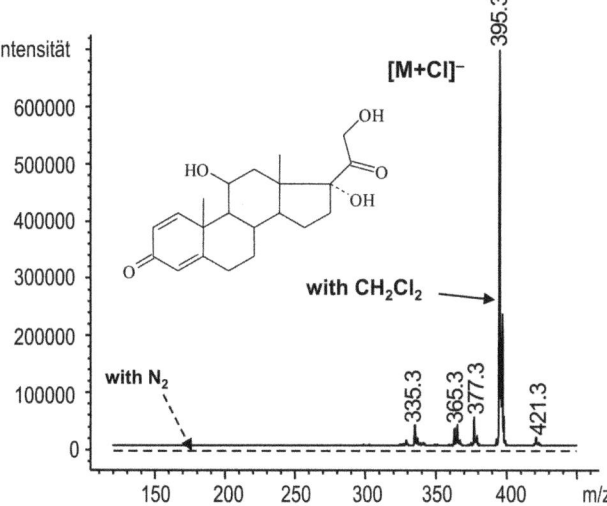

Figure 4.8 APCI adducts of prednisolone.

4.6.3
Summary

The possible additives range from alkali salts and methylene chloride to oxygen. It is up to your own creativity to find your personal "secret recipe." Simply try additives that you know to have high affinities to your analyte's functional groups or which form reactive species in the APCI plasma.

4.7
Tip Number 6

4.7.1
How Can I Enhance Sensitivity of Detection?

Think through all your method parameters step by step. Did you choose the right ionization technique? ESI normally is more sensitive than APCI or APPI. However, if for chromatographic reasons you must select a mobile phase where your analytes are not ionized in solution – you must resort to one of the chemical ionization techniques. But let us assume that you selected an ionization technique suitable for your analyte.

4.7.1.1 Electrospray

The optimal pH value is two units below (positive ion ESI) or above (negative ion ESI) the pK_a value of the compounds to be detected. In this case, more than 99% of your analyte molecules exist as ions in solution. Maybe you have to add acid or base postcolumn in order to achieve these conditions. Some ESI nebulizers generate too large aerosol droplets when spraying 100% aqueous mobile phase. You should then apply more heat – for example by applying "heated drying gas" or "turbo ion spray" – or increase the percentage of organic solvent in the eluent by adding methanol or isopropanol postcolumn. If necessary, combine this with pH value adjustment or the addition of modifiers. Is your background signal significantly stronger than your analyte signal? Then supposedly your ESI detection will suffer from ion suppression. Eliminate the reasons for the high background signal. Try another batch or brand of your HPLC solvents. "HPLC grade" means that this particular solvent is suitable for UV detection, but it does not necessarily mean it is LC/MS compatible. A m/z 102 in positive ion ESI results from previous measurements using triethylamine, m/z 279 is most probably caused by dibutyl phthalate, while a series of peaks at a distance of m/z 44 result from ethoxylated surfactants. If you still use an HPLC column of a larger internal diameter you had better change to smaller column dimensions (3 or 2.1 mm I.D.). Please remember – ESI is a concentration-dependent detection technique. Therefore, the detection sensitivity increases quadratically with the reduction of the column diameter. In addition, the associated lower HPLC flow rate also contributes to the sensitivity enhancement.

4.7.1.2 APCI

In APCI you should check the vaporizer temperature. Two extremes could cause loss insensitivity – too elevated temperatures induce pyrolysis, while too low temperatures result in incomplete evaporation of the analyte. Do you generate sufficient APCI reagent plasma? Try to increase the corona current in small steps. In most cases, this increases not only signal height but also signal stability. Is your HPLC flow rate high enough? As APCI is a mass flow-dependent detector, you gain sensitivity by increasing the column diameter, which increases the HPLC flow rate. Do not set the flow rate too high, though – the optimum for most available APCI ion sources lies in the range of 0.8–1.5 mL/min. If your compounds exhibit lots of OH functional groups, you could also try the additives that have been discussed in the previous section.

4.7.1.3 APPI

The optimal flow rate for APPI is around 0.6 mL/min. If your analyte undergoes direct photoionization (APPI without dopant), then APPI is rather concentration dependent (Lambert–Beer's law), if you use a dopant (acetone or toluene) then APPI behaves like a mass-flow-dependent detector. You should adapt your chromatographic conditions accordingly. Also the vaporizer temperature as well as the capillary or cone voltage influences APPI sensitivity. Unfortunately it is rather impossible to predict the analyte behavior – you have to determine optimum APPI conditions empirically.

4.7.1.4 Optimizing Instrument Parameters in APPI

As the lamp intensity cannot be changed by the operator, the vaporizer temperature, the temperature of the added nitrogen stream and the capillary or cone voltage leave room for optimization. In case you cannot generate sufficient ionization by direct photoionization, you should try with dopant APPI. For that purpose, you add 1–5% v/v of toluene or acetone to the eluent (T-piece right in front of the ion source). Acetone is the best electron donor available for negative ion APPI. Please check on sensitivity by running real samples, not standard samples. In APPI, the absolute intensities often seem to be low. The most important decision criterion, however, is the selectivity of analyte ionization in relation to the chemical background.

4.8
Tip Number 7

4.8.1
No Linear Response and Poor Dynamic Range?

When switching from optical detection techniques to mass spectrometry, you will primarily be bothered by a reduced dynamic range and very often in a nonlinear response. I admit that the dynamic range maybe one to two orders of magnitude lower compared to the use of a diode array detector, mostly it is in the range of 10^3–10^4. However, this limitation typically occurs at high concentrations. The solution of that problem is simple – dilute your sample. In LC/MS, this is normally not a problem, as this detection technique beats any optical detector in sensitivity. When discussing linearity, you should keep in mind that at low concentration we can in most cases observe linear detector response. Again, your first option would be dilution of the sample.

4.8.2
The Reasons

As you have learned already, we generate a high charge density in the spray, in particular when using electrospray. At injected concentrations of about 0.5 mg/mL, the ion source is affected by what is known as space charge limit effect, that is, the charged cloud before the orifice just widens with higher concentration, but the number of ions that reach the mass spectrometer will not increase. What is particularly annoying is that it is mainly the compounds with a high ionization yield that are affected by this phenomenon, also known as "soft clipping." A classical example is shown in Figure 4.9. While o-toluidine exhibits a perfectly linear response in ESI, 3,3′-dimethylbenzidine – being just twice the o-toluidine molecule – shows a quadratic behavior. The duplication of the ionizable functional groups per molecule increases the ionization yield and therefore also limits the dynamic range.

4.8.3
Possible Solutions

Use isotope-labeled internal standards – if possible at a concentration below the "critical" analyte concentration. The space charge effects now influence both the analyte and internal standard. It would be even better if you switched from ESI to APCI or APPI detection. In both techniques the cloud of charged species is less dense. Because of this, the nonlinearity problems are significantly shifted to higher concentrations. The response curves in APCI and APPI mostly are more

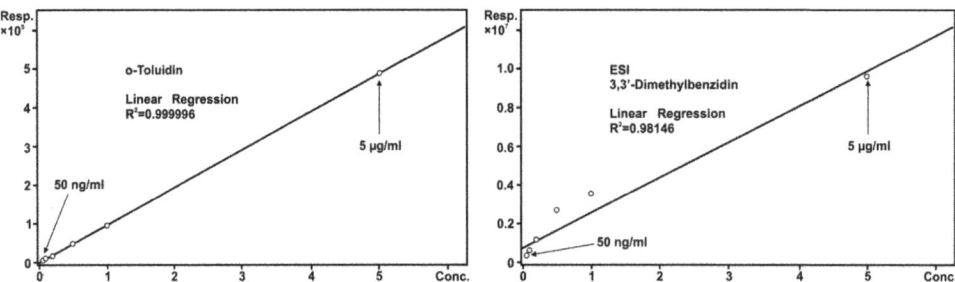

Figure 4.9 ESI calibration curves of o-toluidine and 3,3′-dimethylbenzidine.

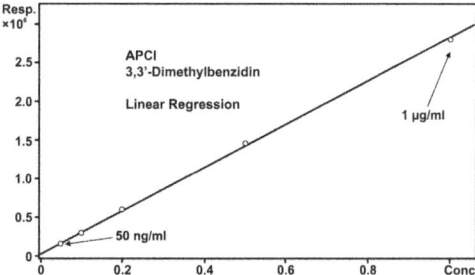

Figure 4.10 APCI calibration curves of o-toluidine and 3,3′-dimethylbenzidine.

linear than in ESI. For example, our case of 3,3′-dimethylbenzidine shows very good linearity in APCI (Figure 4.10).

4.8.4
Summary

High detection sensitivity can be a curse. High ionization yields result in space charge effects in the spray chamber and harm both the dynamic range and linear range of detection, in particular in electrospray mode. Very often you can improve performance by just adapting the concentration range of your samples (dilution). If you have access to an APCI or APPI ion source, you should try to change your method from ESI to these more linear detection techniques. Slight losses in sensitivity compared to ESI will be compensated by a significantly better linearity.

4.9
Tip Number 8

4.9.1
How Much MS^n Do I Need?

There are many reasons to opt for MS^2 or MS^3 (Figure 4.11). Let us focus on decision criteria that are based on analytical reasons. The more complex your injected sample is and the shorter your chromatographic separations, the more specific your detector has to be. This applies to a mass spectrometer the same way as it does to other detection methods. You do not only selectively measure just the molecular ion or one or multiple fragment ions – the high specificity of MS/MS results from the detection of a transition or a reaction from a "precursor" to a fragment. You detect the origin of the signal, not just its existence. This is why in MS/MS you talk about "single reaction monitoring" (SRM) or "multiple reaction monitoring" (MRM). If you monitor full MS/MS spectra instead of MRM transitions, you lose about a factor of 10 in sensitivity. No sensitivity loss happens by using ion trap mass spectrometers.

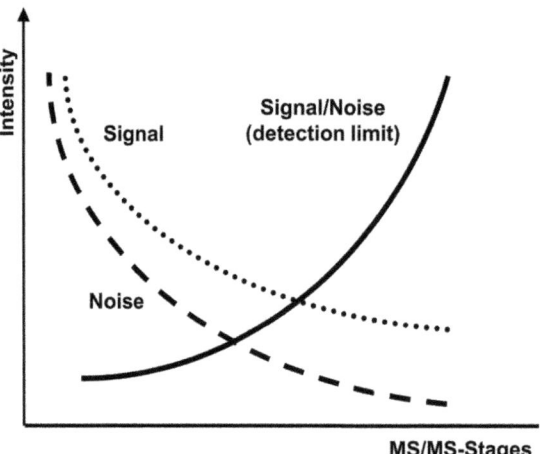

Figure 4.11 Signal/noise as a function of MS^n stages.

4.9.2
Solutions

After having read the previous tips you know already that artifacts (ion suppression, alkali adducts) are caused by the ionization process, regardless of what kind of mass spectrometer you use behind the ion source. The need for a reasonable chromatographic separation of the analytes from the sample matrix increases with the sample complexity. If the analysis time is an important factor in your analytical environment you can hardly work without using an MS/MS-capable mass spectrometer. Even if you use quantitative MS/MS methods that have been optimized in each and every detail, the use of internal standards is highly recommended. These internal standards should undergo the same analytical processes as the analytes (stable isotope labeling).

In case you are just interested in the mass spectrum or the molecular weight in order to confirm a synthesis step, you can even work without any chromatographic separation in flow injection mode (FIA), and you will not need an MS/MS-capable mass spectrometer either. Fragment ion spectra can be obtained by collision-induced dissociation (CID) inside the ion optics – even by using rather simple mass spectrometers (i.e., single quadrupole or time-of-flight instruments). If you follow a very selective sample preparation procedure and/or if you have developed a good chromatographic separation, you can work with high detection sensitivity and selectivity, even without using MS/MS.

4.9.3
Summary

Do not ask yourself "how much MS^n do I need", but "how much effort in chromatographic separation do I need in order to minimize MS-related artifacts".

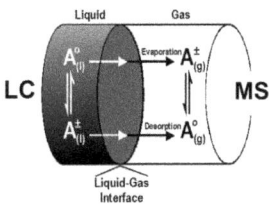

Figure 4.12 Ionization principle at atmospheric pressure.

And finally another five LC/MS tips regarding ionization at atmospheric pressure.

Despite the fact that modern ion source designs are significantly less critical regarding LC flow rates and buffer conditions, I again would like to emphasize some fundamentals of atmospheric pressure ionization (Figure 4.12).

4.9.3.1 Optimization of the Ionization Process

It may appear as a triviality – only analytes in ionic form will be monitored by the mass spectrometer. You remember – in electrospray ionization (ESI) the formation of cations or anions happens in solution. This means that, while entering the ion source, the pH value of the eluent has to be 2 pH units below (positive ion ESI) or above (negative ion ESI) the pK_a value of the analytes in order for 99.9% of the analyte molecules to be protonated or deprotonated. Unfortunately, this is very often in contradiction to the efforts of the chromatographer who would like to develop methods with optimum analyte retention on RP stationary phase materials. Ideally, the analytes have to be in nondissociated form for this kind of separation.

Fortunately, the mass spectrometric detection is so selective that in most cases one can deviate from the ideal situation of baseline-separated LC peaks. So you can run a chromatograph at a slightly acidic pH while still being able to detect analyte anions in negative ion electrospray, but at suboptimal detection sensitivity. In the same way, you can detect analyte cations in positive ion electrospray after having adjusted an alkaline pH by adding ammonium hydroxide to the mobile phase. Again I would like to emphasize that all that typically comes with a drastic loss in sensitivity.

In case of an incompatibility between the optimal pH values for either chromatographic separation or ionization via acid/base equilibrium, there are several possibilities to overcome that problem. If you have an additional HPLC pump available, you can add acid or base postcolumn in order to shift the pH value to a range which is optimal for electrospray ionization. This can even mean an inversion of the pH. It is important that the added buffer does not dilute the eluent too much (electrospray is a concentration-dependent detection technique) and that the analytes do not precipitate.

Besides the addition of acid or base after the separation column, many analytes are cationized or anionized by the formation of adducts. For example, compounds which contain OH functional groups tend to form adducts with alkali cations. The recommended alkali concentration is below 1 mmol/L (i.e., sodium acetate) in order to avoid ion suppression effects. It is important to achieve the cation adduct formation by postcolumn addition and not at all by addition to the HPLC buffer. This would cause a long-lasting contamination of the HPLC system. Carbohydrates can be detected as chloride adducts in negative ion electrospray by adding 50 mM HCl (2 mmol/L in the ESI ion source).

Alternatively you can change to another ionization technique where the cationization or anionization takes place in the gas phase. These are chemical ionization (atmospheric pressure chemical ionization or APCI) and photoionization (atmospheric pressure photo ionization or APPI) at atmospheric pressure.

With those ion sources the eluent is entirely evaporated before the ionization of the analytes. This has the disadvantage of some thermal stress to the analytes because the mobile phase is sprayed into a heated ceramic or quartz cartridge. Although these cartridges are heated up to 450 °C, the effective temperature inside the cartridge is in the range of only 120–150 °C. However, thermolabile compounds like sugars or peptides do not survive this evaporation process. Also quite common is the elimination of water (steroids) or CO_2 (carboxylic acids).

The advantage of these two ionization modes is their lower tendency to interfere with the sample matrix, unlike electrospray.

4.9.3.2 Lost LC/MS Peaks

Let us assume that your sample introduction, that is, by the use of an automated injector, functions flawlessly. You then inject an analyte amount which actually should give a reasonable signal intensity. These amounts are in the range of ictograms to nanograms, depending on the type of mass spectrometer and its operating mode. Assuming that you see an injection peak in the UV or MS signal, the sample supposedly has been transferred completely onto the HPLC column. However, the mass spectrometer monitors peaks but your expected analyte peak has "disappeared". Actually, there can be only three reasons if your MS signal disappears in electrospray.

Mobile phase pH at the edge of the optimum range: in order to obtain a good chromatographic separation, the pH value of the mobile phase is very often adjusted to a suboptimal value for the subsequent ionization in electrospray. Furthermore, it is possible that the pH of the mobile phase has changed due to long-term storage in the HPLC system. Just measure the pH of a 0.1% aqueous formic acid or acetic acid solution directly after preparation and after several days of storage. In most HPLC systems, the solvent bottles are open to the laboratory atmosphere and therefore lead to a slow pH increase in the example mentioned above. Even more unstable is the pH of an ammonium hydroxide solution for negative ion electrospray. You can easily compensate this kind of pH drift by postcolumn addition of freshly prepared acid or base at reasonable concentrations.

Ion pairing agents in the HPLC system: ion pairing agents are poison to the electrospray ionization. The analyte ions which actually should be released during the electrospray process are blocked in the ion pair and therefore do not result in an MS signal. Should you not be able to avoid ion pair chromatography due to strong analyte polarity, you must choose as weak as possible ion pair reagents (i.e., perfluorinated aliphatic carboxylic acids or aliphatic amines) and better choose

APCI as an ionization method. However, the unintended presence of ion pairing agents in your HPLC system may also suppress your MS signal by 100%. A typical example is seen in Figure 4.6 where the negative influence of TFA on the LSD metabolite LAMPA is demonstrated. Independent from the pH value, TFA completely suppresses the LAMPA signal. As you can see, even with minimal differences in chemical structure, it is extremely different to predict the tendency of analytes to form ion pairs.

Even minimal residues of ion pairing agents in solvent bottles, tubing, vacuum degasser, or the chromatographic column exhibit negative influence on the ionization in electrospray mode. In case of uncertainty, please speak to your colleagues whether by accident ion pairing agents found their way into the jointly used HPLC system ("…I only used 0.001% aqueous TFA…"). If this happened, take out all contaminated tubing, bottles, columns, bypass the module with the worst memory effect – the vacuum degasser – and flush the system several hours to days (!). If you know the chemical structure of the contaminant, please monitor the decay of its signal (i.e., m/z 113 for the TFA anion).

Ion suppression by the sample matrix or sample contaminants: the interference of the ionization of an analyte with a disturbing component is widely called "ion suppression", even if the signal gets enhanced instead of suppressed. Most prone to this effect is of course electrospray, but also more rarely APCI and APPI. This interference of eluting compounds is a complex issue and will be extensively discussed later.

4.9.3.3 How Clean Does an LC/MS Ion Source Have to Be?

Those users of LC/MS with some experience in mass spectrometry tend to polish ion source and ion optics before important sample series. Of course, you will achieve maximum signal intensities in API LC/MS ionization techniques if you completely eliminated all residues from the ion source surfaces. The electrical fields that are necessary to form and/or transport the ionic analytes before they reach the mass analyzer will be influenced by the degree of contamination with nonvolatile sample constituents. On the other hand, experienced LC/MS users know that the long-term stability of an LC/MS system will be increased if you intentionally contaminate the spray chamber slightly. This is of particular importance if you want/have to quantify by using external calibration. With the most commonly used ion sources, it is enough to inject sample matrix 10–20 times in order to achieve an almost constant analyte response – after an initial exponential drop (Figure 4.13).

However, after long series of heavy sample matrix load, the response will significantly drop. This is strongly dependent on the instrument type and brand used. Those regions of the ion source that are exposed to atmospheric pressure are easily regenerated by just flushing with solvents. If you are unable to get your signal back by this, then it is very likely that the ion optics (ion transfer capillary, skimmer, cone, ion guide) are seriously contaminated.

Before you now start to vent your LC/MS system in order to clean the interior – let me show you a trick on how you can recover the system for a few

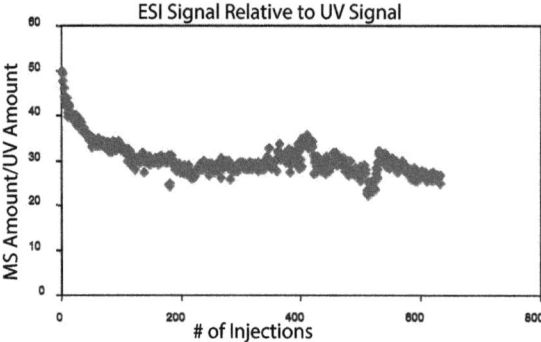

Figure 4.13 Drop of signal intensity after extreme matrix stress of an ion source.

runs. Often layers on the surfaces of the ion optics (skimmer, cone) cause electrostatic charging or an unwanted hysteresis when applying the operating voltages. Change ion polarity for several minutes – then all DC operated elements will become discharged. This can be even included in your acquisition method in order to increase the uptime of your instrument. Just change ion polarity at the end of your chromatographic run while you re-equilibrate the chromatographic column. However, those elements in the ion optics that are operated at AC voltages (quadrupole/hexapole/octopole) will not be revitalized from contamination by that procedure. Then the only solution will be venting of the system and cleaning – unfortunately. The best prevention against frequent instrument cleaning is the diversion of the HPLC void volume to waste. Almost all nonvolatile components in your sample will elute in the front of your chromatogram. Therefore, please consequently use the diverter valves of most of today's LC/MS systems.

4.9.3.4 Ion Suppression

The interference of co-eluting compounds with the ionization of the analytes of interest is the most frequent reason for insufficient response and reproducibility in LC/MS. The phrase "ion suppression" was introduced in the mid 1990s and describes the phenomenon only partially, as an enhancement of the signal can also occur. In particular, "endangered" are methods with little sample preparation and/or little chromatographic separation. Detection methods that monitor just single ions or MS/MS transitions (SIM, MRM) are quasi blind to co-eluting compounds, while the detection of complete mass spectra allows to recognize massive co-elution and to take corrective action.

After the initial euphoria about the high selectivity of LC/MS-based methods had calmed down it has come to some disillusionment. It has been recognized immediately that compounds that are invisible in MS detection are not always "inactive." The matrix effect takes place solely in the ion source and, therefore, is independent from the selectivity of the subsequent MS(/MS) detection. Ion suppression happens in all ionization modes – electrospray, APCI and APPI. However, the strength of the effect may depend on the design of the ion source. Please

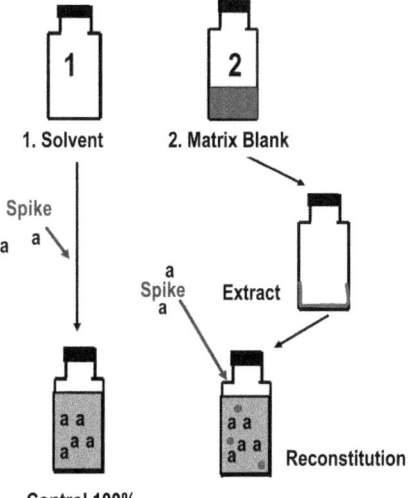

Figure 4.14 Comparison of solvent standard with matrix spike.

do not search for the reason not only in your sample as matrix effects may be also caused from exogenic materials. In particular in gradient HPLC methods, organic substances which are dissolved in the mobile phase or which bleed from parts of your HPLC system elute with your analytes of interest in the gradient and may massively influence the signal strength. Unfortunately "HPLC grade" solvents are not necessarily suitable for the use in LC/MS. It is highly recommended to test solvents of different suppliers as well as various qualities and batches and to then decide based on the lowest background signal. LC/MS systems also disclose rapidly if you had contaminated your mobile phase bottles with phthalates from plastic caps or with surfactants from the laboratory dish washer.

The matrix effect can be produced by various processes – hindering of the analyte ion release in electrospray, that is, by alkali or phosphate, micelle formation by surfactants, formation of ion pairs or even precipitation of the analytes. It is important that you specifically try during method development to recognize and exclude ion suppression or matrix effects.

In order to evaluate a matrix effect you have the choice of various experimental approaches. The easiest way is to compare a "solvent standard" with a "matrix spike" with each other. Figure 4.14 describes the corresponding steps [1].

Ideally the sample matrix does not have any influence on the analyte signal. It is recommended to test with matrix blanks of different origin to exclude random results. The approach described in Figure 4.14 can be extended with little effort for additional determination of the method recovery and the extraction yield. The required steps are described in Figure 4.15.

The comparison of steps 1 and 3, the solvent standard without extraction and the extracted/reconstituted matrix spikes results in the method recovery. From

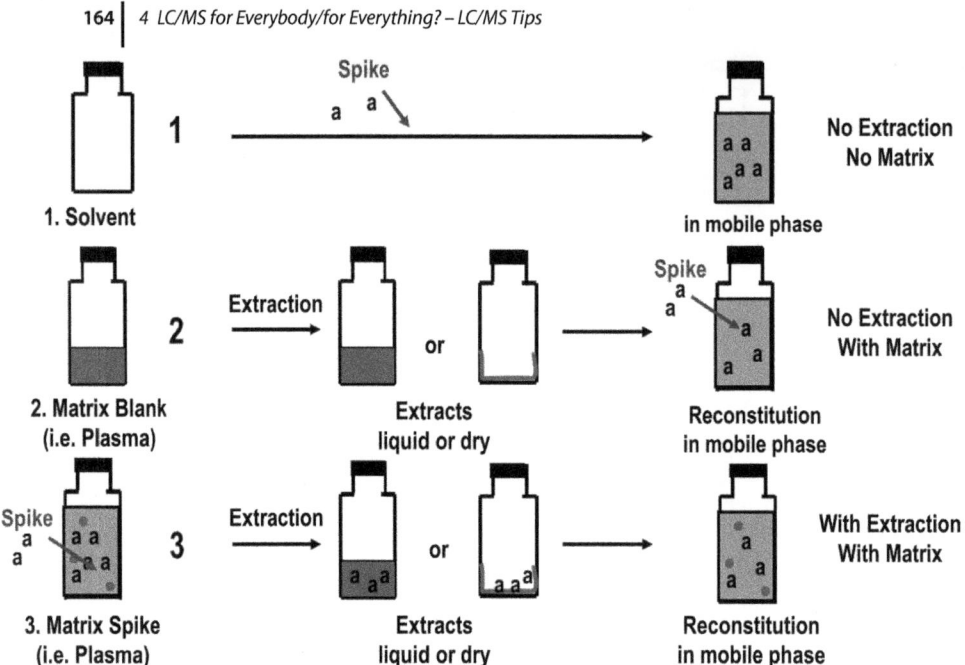

Figure 4.15 Extended evaluation of the matrix effect.

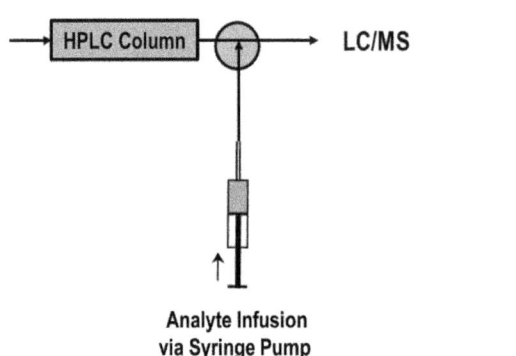

Figure 4.16 Method optimization by analyte infusion.

steps 1 and 2 the matrix effect can be determined. For the calculation of the extraction yield, steps 2 and 3 are compared with each other (Figure 4.15).

For the optimization of LC/MS parameters, it has been proven to continuously introduce the analyte into the ion source (Figure 4.16). This syringe infusion generates a constant analyte signal. In order to create ionization conditions identical to the intended LC/MS analysis, the analyte is usually infused into the eluent via a T-piece behind the separation column (Figure 4.16).

With that experimental setup, it is quite easy to evaluate the matrix effect. Instead of introducing the analyte into the neat mobile phase, we simultaneously run a chromatographic separation of a matrix blank. Without any matrix effect,

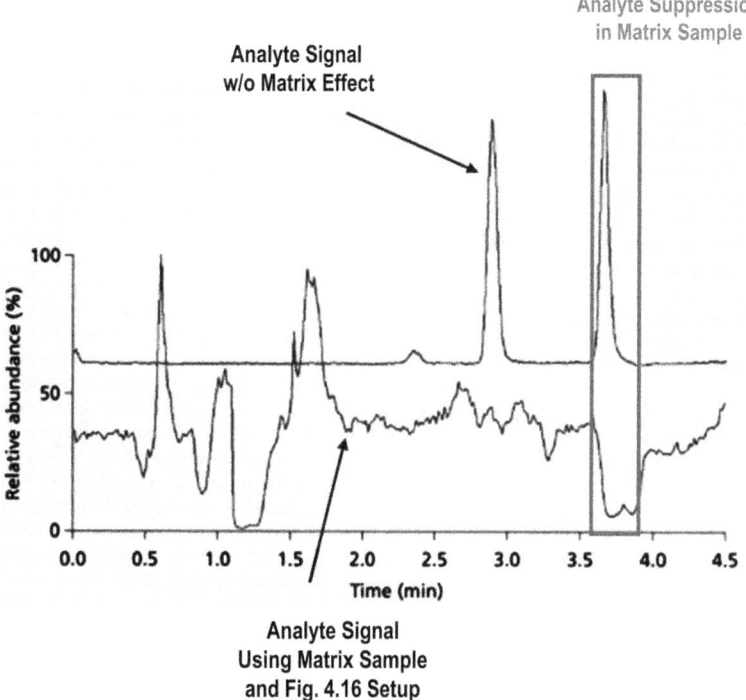

Figure 4.17 Evaluation of the matrix effect by analyte infusion.

we obtain a constant analyte signal. By infusion of the analyte into the eluent with its various chromatographic peaks, we will observe an enhancement or a suppression of the analyte signal in retention time regions of high matrix content. In the final acquisition method, the analytes should elute at retention times which exhibit neither enhancement nor suppression. Figure 4.17 shows an overlay of the infusion signal of the analyte with a solvent standard.

Whatever approach you chose to evaluate the ion suppression or matrix effect – you will have to vary your analytical method. The presumably simplest step is the variation of your chromatographic system. Change the gradient slope and/or change the mobile phase (acetonitrile vs. methanol, etc.). If possible, change your extraction procedure (SPE vs. liquid/liquid). Try to change the ionization method. Matrix effects are less severe in APCI and APPI than in electrospray. If you have access to instruments of different ion source designs, change to a different LC/MS instrument. Use isotope-labeled internal standards for method calibration. The advantages are the same extraction yield as your analytes, the same ionization yields, and the same chromatographic retention times (^{13}C better than ^{2}H). Even when matrix effects should occur, they will act on both analytes and the internal standards in the same way.

4.9.3.5 Ammonium Fluoride – a Quite Unusual Buffer Additive

Volatile ammonium salts of organic acids are typical buffer additives in LC/MS. Besides the "classical" ones such as ammonium formate and acetate, ammonium fluoride is a kind of secret tip – in particular for electrospray LC/MS. Unlike common buffer concentrations of 5–10 mmol/L, ammonium fluoride requires even ten times lower concentrations. In positive ESI, the formation of sodium adducts is reduced by ammonium fluoride. In particular for MS/MS methods, it is important to avoid alkali adduct formation, as the positive charge is eliminated together with the alkali cation and therefore is no longer available for the remaining molecule or molecular fragment. But also in negative ESI, ammonium fluoride tends to enhance ionization yields, such as for the signal response of organic acids or sugar phosphates. That effect can also be observed with other compound classes, such as steroids. In case of negative ESI signal reduction after having added ammonium fluoride, most likely fluoride adducts had been formed. One typical example are oligonucleotides.

4.10
Need More Help?

What else can you do in order to successfully use LC/MS? Read the manuals that come with your LC/MS system. Here you will find hints on how to enter method parameters, optimization, and tuning. In general, the instrument manufacturers provide default methods with their instruments, which are well suited as starting points for your own method development.

It is false economy not to visit the operator training courses your instrument manufacturer offers. There your LC/MS system's features will be explained to you. "Classroom training" will give you the opportunity to meet other users with similar interests. After the training course you should keep in touch with the other participants. If you want to brush up on theory – there are manufacturer-independent training programs where you could become familiar with the fundamentals of LC/MS.

Rather than re-inventing the wheel you should ask colleagues or your manufacturer's application chemists for advice. Application chemists are confronted with a wide range of analytical problems on a daily basis. In most cases, they are able to help you with your questions. Also for troubleshooting they are the right person to contact.

Read LC/MS-relevant literature. Using Internet-based search engines you can easily find suitable information. On the websites of the LC/MS manufacturers you can find a variety of application examples. Methods which have been reported using your type of instrument are easy to transfer to your setup. However, please consider the different instrument designs of the different manufacturers – ESI sprayers, for example, can have orthogonal or off-axis geometry, heating could be realized via heated nitrogen or a heated ion transfer capillary, and much more. Increasing a "cone voltage" may have the same effect as increasing a "fragmentor voltage" – more and more fragmentation of the molecular ions. However, the absolute voltage settings may differ significantly. This is just one example on how different approaches (instrument parameters) lead to the same result (mass spectrum).

The "take home message" is that LC/MS(/MS) is simple in principle. Trouble is caused by the ionization chemistry and by the fragmentation behavior of the analytes. Do not give up all too soon – even the so-called "experts" with 5 years' experience had to go through those 5 years first.

References

Literature

1 Choi, B.K., Hercules, D.M., and Gusev, A.I., (2001) Effect of liquid chromatography separation of complex matrices on liquid chromatography–tandem mass spectrometry signal suppression. *J. Chromatogr. A*, **907**(1–2), 337–342.

Further Reading

2 Fenn, J.B., Mann, M., Meng, C.K., Wong, S.F., and Whitehouse, C.M. (1989) Electrospray ionization for mass spectrometry of large biomolecules. *Science*, **246** (4926), 64–71.
3 Nohmi, T. and Fenn, J.B. (1992) Electrospray mass spectrometry of poly(ethylene glycols) with molecular weights up to five million. *J. Am. Chem. Soc.*, **114** (9), 3241–3246.
4 Labowsky, M.J., Whitehouse, C.M., and Fenn, J.B. (1993) Three-dimensional deconvolution of multiply charged spectra. *Rapid Commun. Mass Spectrom.*, **7**, 71.
5 Fenn, J.B. (1993) Ion formation from charged droplets: Roles of geometry, energy and time. *J. Am. Soc. Mass Spectrom.*, **4**, 524.
6 Bruins, A.P. (1991) Mass spectrometry with ion sources operating at atmospheric pressure. *Mass Spectrom. Rev.*, **10**, 53–77.
7 Niessen, W.M.A. and Tinke, A.P. (1995) Liquid chromatography–mass spectrometry. General principles and instrumentation. *J. Chromatogr. A*, **703**, 37–57.
8 Tomer, K.B., Moseley, M.A., Deterding, L.J., and Parker, C.E. (1994) Capillary liquid chromatography mass spectrometry. *Mass Spectrom. Rev.*, **13**, 431–457.
9 Wachs, T., Conboy, J.C., Garicia, F., and Henion, J.D. (1991) Liquid chromatography–mass spectrometry and related techniques via atmospheric pressure ionization. *J. Chromatogr. Sci.*, **29**, 357–366.

Internet

10 www.spectroscopynow.com.
11 www.asms.org.
12 www.dgms.de.
13 www.lcms.com.
14 www.ionsource.com.
15 http://masspec.scripps.edu.

Part III
User Reports

5
LC Coupled to MS – a User Report

A. Muller and A. Hofmann

In comparison to coupling liquid chromatography (LC) to UV, coupling LC to mass spectrometry (MS) requires important adaptations. Water–methanol or water–acetonitrile gradients are often applied in reversed-phase (RP) chromatography. Only volatile buffers should be used in order to prevent contamination of the MS instrument by nonvolatile salts. For example, formic acid or acetic acid can be used for acidic conditions and ammonia can be used for alkaline conditions. If samples contain nonvolatile salts, the LC flow can be directed to waste at the beginning of the analysis to prevent contamination of the MS instrument. A two-dimensional LC system with an enrichment column is another option to analyze samples containing nonvolatile salts. Matrix effects can reduce signal to noise ratios of analytes in complex biological samples. Adding heavy isotope labeled standards to samples allows to take losses during extraction and matrix effects into account. ^{2}H, ^{13}C, and ^{15}N are often used to isotopically label standards. In contrast to ^{2}H-labeled standards that can show small retention time shifts, ^{13}C- and ^{15}N-labeled standards elute at the same time as the analytes.

The most suitable LC system can be selected based on the required sensitivity and the desired analysis time. The MS signal intensity is proportional to the concentration of the analyte. Nano-LC systems are mainly used to achieve highest sensitivity. High sensitivity is especially critical for the analysis of low abundant endogenous molecules, for example, in proteomics experiments. In proteomics experiments, chromatographic columns have often an inner diameter of only 75 µm or lower. The low solvent flow rate (e.g., 250 nl/min) in nano-LC/MS methods can lead to an analysis time of 1 h and more, but allows the most sensitive approach to detect endogenous molecules. Higher LC flow rates (400–600 µl/min) are usually applied for the detection and quantification of analytes with higher concentrations. Chromatography columns with a 1 mm or 2.1 mm inner diameter are often used for U(H)PLC applications that allow to complete an analysis in only a few minutes.

The method of ionization is chosen based on analyte properties and the LC flow rate. ESI (electro spray ionization) is suitable for a very broad spectrum of analytes and LC flow rates. Hydrophilic as well as hydrophobic substances can be analyzed well with ESI. APCI (atmospheric pressure chemical ionization) and

The HPLC-MS Handbook for Practitioners, First Edition. Edited by S. Kromidas.
© 2017 WILEY-VCH Verlag GmbH & Co. KGaA. Published 2017 by WILEY-VCH Verlag GmbH & Co. KGaA.

APPI (atmospheric pressure photo ionization) are mainly applied for the analysis of very hydrophobic substances and require high LC flow rates. In general, the ion source temperature and the gas flow rates need to be increased with increasing solvent flow rates. Recommended parameters for different LC flow rates can be found in the manual of the ion source. These recommended parameters often represent a good starting point for your own optimization. In addition to parameters that depend on the solvent flow rate, analyte-dependent parameters of the ion source need to be optimized. Analyte-dependent parameters can be optimized by direct infusion of a pure analyte solution with a syringe pump, in order to adjust all parameters for a maximal signal to noise ratio. Combining the flow of the syringe pump via a T-piece with the LC flow allows conditions close to the final analysis conditions to be simulated. Relatively high concentrations of analyte have to be used for direct infusion, which can lead to background signals of the analyte. Different instrument parameters can also be directly varied during the LC/MS analysis, due to the short analysis time of U(H)PLC systems and the fast scan speed of modern MS instruments. A lower concentration of analyte can be used for the variation of parameters during the LC/MS analysis compared to direct infusion. Many steps of the optimization can automatically be done by the instrument software or platform independent software. However, automatically determined parameters should always be checked for plausibility. Orifices of MS instruments have been enlarged over the last years in order to increase ion transmission. Not only more ions but also more neutral particles enter the MS instrument through the enlarged orifice, therefore, strong roughing and turbo pumps are necessary to maintain the required vacuum. Furthermore, the geometry of the ion optics were changed to efficiently separate ions from neutral particles, for example, by using the StepWave™, iFunnel, or electrodynamic ion funnel technology. The improved ion transmission of modern MS instruments results in higher signal to noise ratios, more robust methods, and facilitated optimization of instrument parameters.

Mass accuracy, mass resolution, scan speed, sensitivity, and many other parameters of MS instruments can differ greatly. Some applications may only be feasible with one specific type of MS instrument. Basic MS instruments, such as single quadrupole, ion trap or TOF (time of flight) instruments, can be used to analyze samples with low complexity. Hybrid MS instruments, combining two mass analyzers, are often used to analyze complex biological samples. Hybrid MS instruments with high mass accuracy and mass resolution, for example, Q-TOF, TOF-TOF, ion trap-Orbitrap, or Q-Orbitrap instruments, allow the identification of unknown substances. Triple quadrupole MS instruments with high scan speed and excellent sensitivity are often applied for quantitative analyses. A triple quadrupole instrument consists of three quadrupoles arranged one after the other (Figure 5.1a). The first quadrupole is used as a mass filter for the ionized, intact analyte. The ionized, intact analyte is also called parent ion. The parent ion is then fragmented in the second quadrupole by collision induced dissociation. Subsequently, the third quadrupole filters for a specific fragment of the parent ion, also

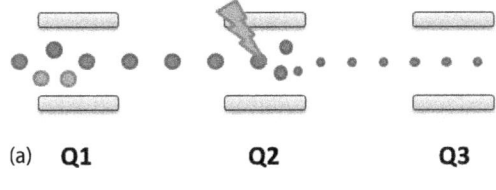
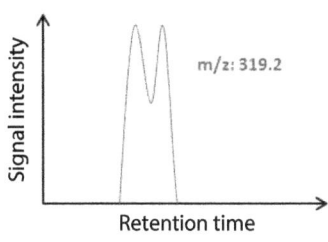
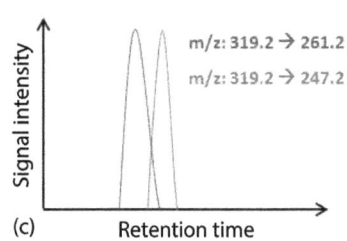

Figure 5.1 (a) Schematic representation of the quadrupoles of a triple quadrupole instrument. Only two quadrupole rods of the four quadrupole rods are shown for each quadrupole. The first quadrupole (Q1) filters for the precursor ion, the precursor ion is fragmented in Q2 by collision induced dissociation and Q3 filters for a specific daughter ion. (b) Isomers of hydroxyeicosatetraenoic acid (HETE). (c) Schematic representation of ion traces by measuring just the precursor ions or daughter ions of HETE isomers. Reproduced with kind permission of Novartis.

called daughter ion. Assays on triple quadrupole instruments are very sensitive and selective due to the double filtering in the first and third quadrupole.

For example, isomers of the hydroxyeicosatetraenoic acid are difficult to separate chromatographically and a simple mass analysis (Figure 5.1c) shows overlapping peaks for the parent ions. However, a triple quadrupole instrument enables to measure specific daughter ions of the two isomers and the daughter ion traces show no interferences between the two isomers [1].

Ion pairing reagents, such as trifluoroacetic acid (Figure 5.1b) for acidic conditions or triethylamine for alkaline conditions, can be used for the separation of hydrophilic analytes by RP chromatography. In general, ion pairing reagents often lead to a reduced signal-to-noise ratio and to high background signals when the polarity is switched. Ion chromatography (IC) is a reliable alternative to separate very hydrophilic, charged molecules which cannot be readily analyzed by RP chromatography. Analytes are separated based on their charge and size by IC. In contrast to silica-based RP columns, the stationary phases of IC columns are polymer-based. Therefore, IC columns are very stable under alkaline conditions. Potassium hydroxide is often used for anion exchange chromatography and the eluting strength is directly proportional to the potassium hydroxide concentration. Methanesulfonic acid is often used for cation exchange chromatography.

An example of an anion exchange IC coupled to a triple quadrupole MS is described in the following section. A potassium hydroxide gradient is produced by the eluent generator and analytes are separated on the IC column. Thereafter, potassium ions are exchanged by hydronium ions in the electrochemical suppressor. Usually, IC is coupled to a conductivity detector, which determines the conductivity of the solution. A conductivity detector is relatively insensitive and not well suited for biological applications, for example, in the field of metabolomics. Furthermore, the conductivity is not selective so that analytes with the same retention time cannot be distinguished. In contrast, a triple quadrupole instrument enables a very sensitive and selective detection of different analytes. Without electrochemical suppressor, potassium ions would lead to a strong suppression of analyte signals and contamination of the MS instrument. The electrochemical suppressor generates an aqueous solution with a low salt concentration of a few microsiemens. The efficiency to generate negative ions can be increased by adding an organic solvent, such as methanol, via a T-piece to the aqueous solution. A Dionex ICS-3000 IC system, an ESI ion source and an AB SCIEX QTrap 5500 MS instrument was used for the here described analysis. The triple quadrupole instrument allows a very sensitive and selective detection of analytes via MS/MS experiments (Figure 5.2).

Conditions of ion chromatography

Eluent generator	EGC III KOH
Enrichment column	Ion Pac AG20 2×50 mm
Analytical column	Ion Pac AS20 2×250 mm
Column temperature	3 °C
IC pump	Isocratic, 250 µl/min
Suppressor electric current	62 mA
Loop volume	2 µl
Methanol flow rate	50 µl/min

Gradient generator

Time (min)	[OH$^-$] (mM)
0	10
7.5	45
17.5	48
18.5	100
22.5	100
22.6	10
25	10

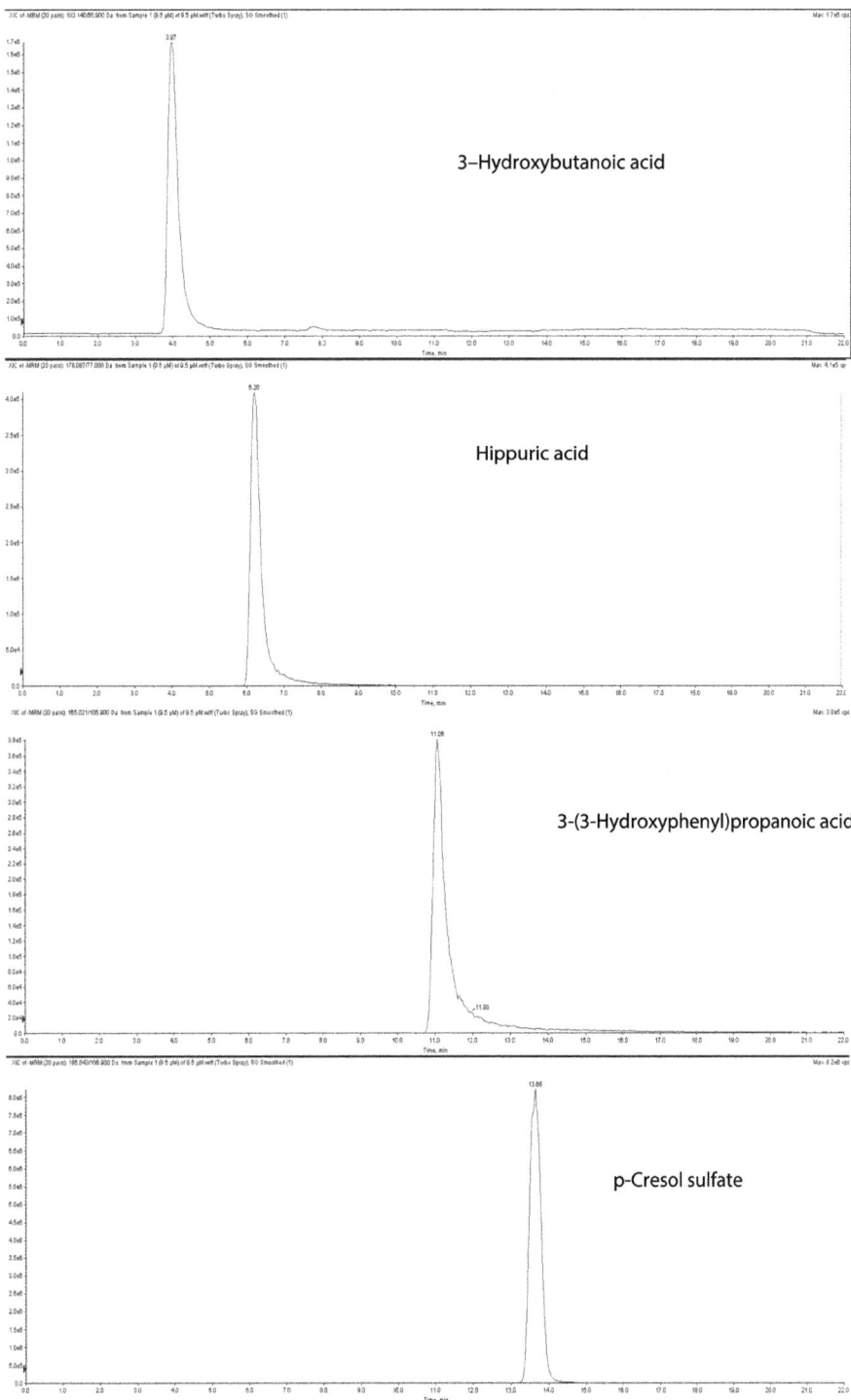

Figure 5.2 Chromatographic separation of four hydrophilic molecules by IC and detection with a triple quadrupole MS instrument. Reproduced with kind permission of Novartis.

Transitions

Analyte	Q1 mass (Da)	Q3 mass (Da)
3-Hydroxybutanoic acid	103.1	58.9
Hippuric acid	178.1	134.1
3-(3-Hydroxyphenyl)propanoic acid	165.0	105.9
p-Cresol sulfate	186.8	106.9

References

1 Dumlao, D.S., Buczynski, M.W., Norris, P.C., Harkewicz, R., and Dennis, E.A. (2011) High-throughput lipidomic analysis of fatty acid derived eicosanoids and N-acylethanolamines. *Biochim. Biophys. Acta*, **1811** (11), 724–736, doi:10.1016/j.bbalip.2011.06.005.

6
Problem Solving with HPLC/MS – a Practical View from Practitioners
E. Fleischer

6.1
Introduction and Scope

LC/MS coupling has proved to be a powerful tool in the course of our everyday research environment and has resulted in a wide range of applications: elucidation of chemical structures, stability testing, decomposition studies, product tracking, reaction kinetics, and evaluation of formulations. Sample preparation was found to be the key to success facing various tasks, besides choosing an appropriate column and eluent. In all these tasks, very precise observation, patience, and last but not least imagination is necessary to draw the correct inference and subsequently to take the best possible action. Interpreting valid data in the wrong way or ignoring small details can be very time consuming.

In the following example, a decomposition product with the mass (M−H) = 272.1 had a nearly identical retention time as the desired chemical product during the synthesis method development. Assuming fragmentation in the mass analyzer, the identical behavior of the decomposition product could lead to the conclusion that the desired product was formed during the method development. In fact, the signal consisted of the mostly undesired decomposition product and only of traces of the target molecule. Due to its structure, it is ionized in ESI-positive and ESI-negative mode very efficiently. Sometimes it is recommended to use positive as well as negative voltage. This practice may lead to an improvement in structure elucidation. Figure 6.1a shows the chromatogram of an impurity that was obtained during a stability test after 40 min with ESI-pos compared to ESI-neg in Figure 6.1b. Only the ESI-positive mode reveals the component at 2.33 min.

The special software function "display mass" (or "extract mass") allows the user to distinguish between interfering signals (e.g., solvent impurities) and relevant molecular masses. Most of our experiments relied on electrospray ionization (ESI) as the method of choice but also API (atmospheric pressure ionization) as well as MALDI (matrix-assisted laser desorption ionization) were used. In special cases like enoxaparin, DAD and also ELSD (evaporative light scattering detector) were applied. Light scattering detection is not used as a routine technique, but the combination of MS-DAD-ELSD (connection via a "T"-splitter enables par-

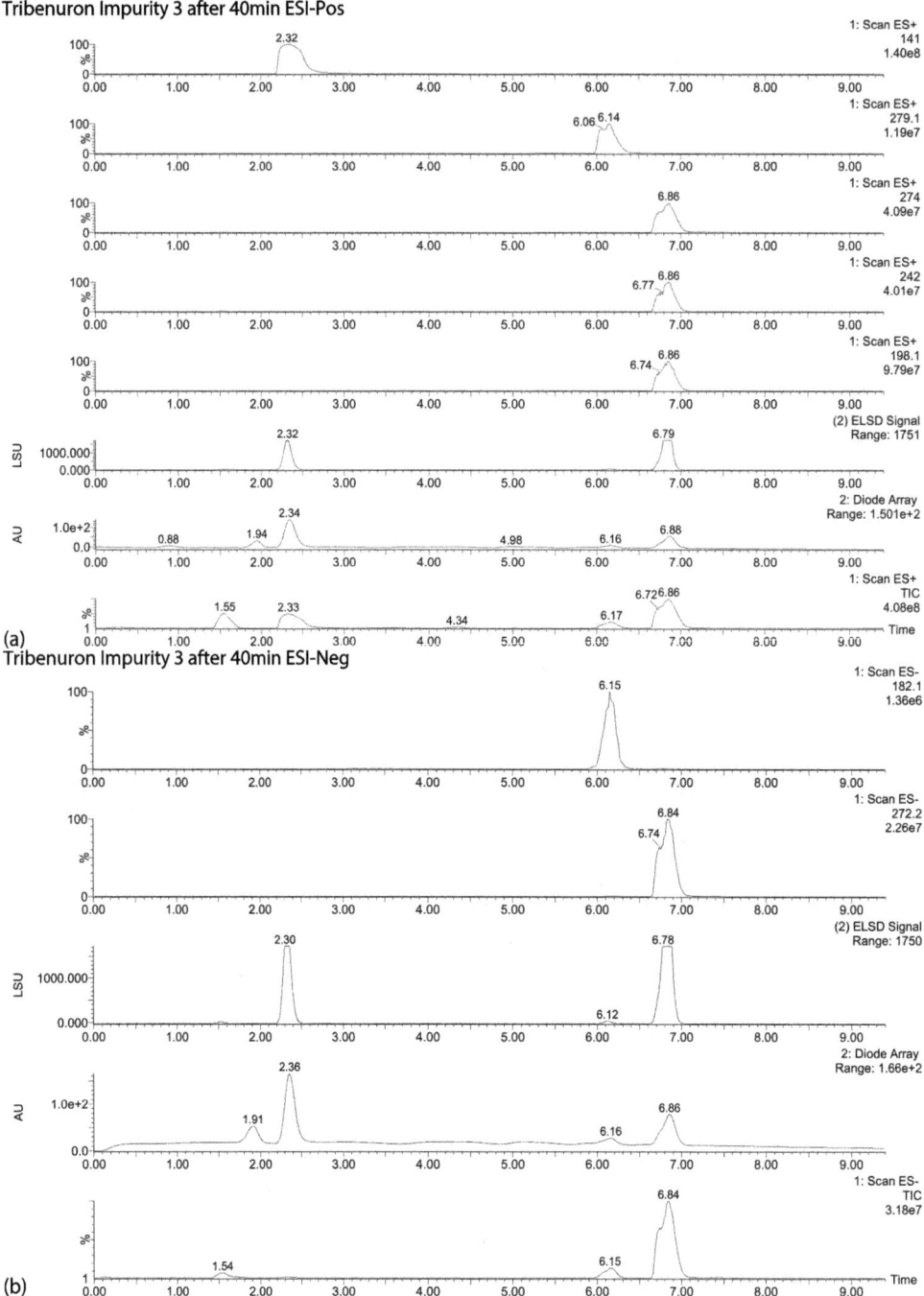

Figure 6.1 Wrong choice of ionization mode can lead to incorrect conclusions in stability tests, for details see text. Reproduced with kind permission of MCC.

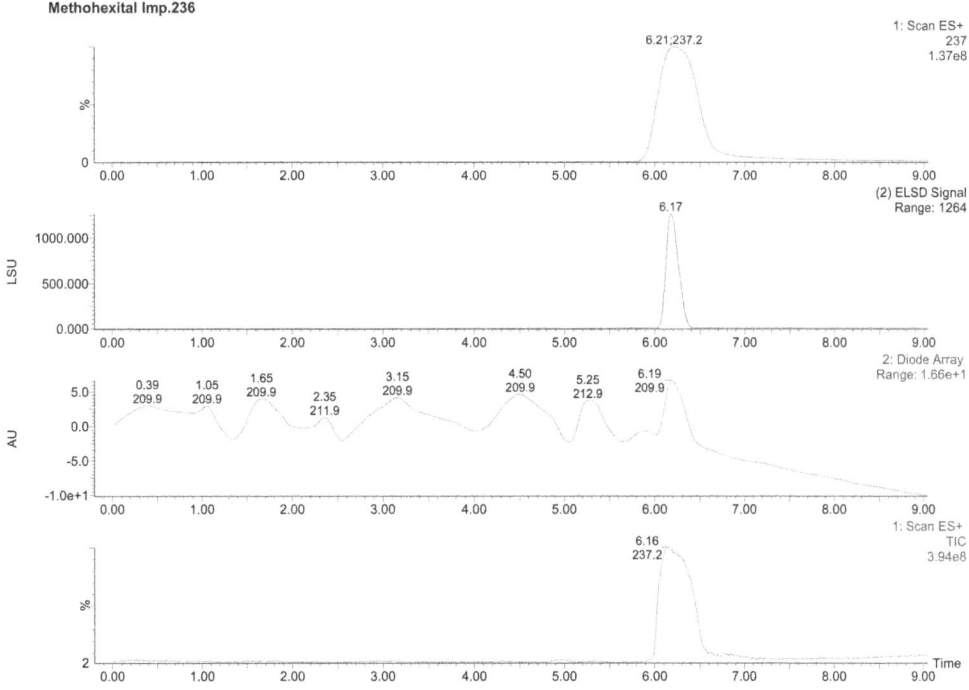

Figure 6.2 The comparison of chromatograms with MS, DAD, and ELSD detection increases the amount of information and, thereby, the accuracy of statements regarding structure and purity. Here: Methohexital Imp. 236. Reproduced with kind permission of MCC.

allel measurements) facilitates significantly the detection, structure elucidation, and discrimination of ambiguities (Figures 6.2 and 6.3).

In most of the cases, particle-based HPLC columns with average particle sizes between 3 and 5 µm and also monolithic material like Chromolith Performance were used. The column diameters ranged from 2.1–4.6 mm and the lengths from 50 to 150 mm. The flow rates were kept between 0.5 and 1.2 mL/min. The high permeability and porosity in the case of Chromolith material results in a low counter pressure, thereby, enabling higher flow rates without sacrificing separation efficiency. We found that this stationary phase is highly suitable for polar molecules with molecular weights of 500 g/mol and more. For example, coenzyme A and its derivatives could only be separated using Chromolith columns. These molecules tend to get "stuck" on particle-based phases and cannot be eluted even with 100% organic modifier over a long period of time. Furthermore, Chromolith columns were used for the analysis of ingredients in cosmetics like glycerophospholipids and phosphatidylcholine derivatives.

About 98% of analytic and preparative tasks in medicinal and pharmaceutical chemistry could be solved with four types of columns: Atlantis T3, Xbridge BEH C18, CSH-XSelect from Waters and Chromolith columns from Merck or Phenomenex. Our personal recommendation because of a broad working range for

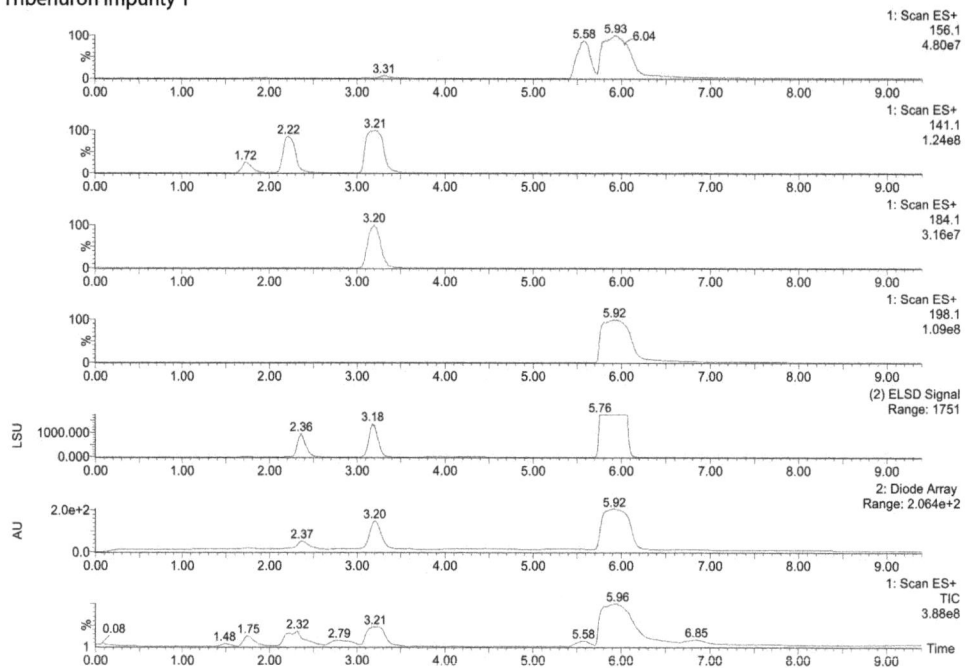

Figure 6.3 The comparison of chromatograms with MS, DAD, and ELSD detection increases the amount of information and, thereby, the accuracy of statements regarding structure and purity. Here: Tribenuron Impurity 1. Reproduced with kind permission of MCC.

polar and nonpolar solvents as well as an excellent pH stability, are the Waters CSH-XSelect columns. We used linear gradient elution with acetonitrile with 0.1% formic/TFA acid or 0.1–0.2% ammonia.

Important observations

In some cases, during preparative chromatography, crystallization of the analytes in the instrument's capillary system occurs. A mixture of acetonitrile/methanol 1 : 1 proved to be an effective inhibitor of clogging when the nature of the analytes and the separation method allows. Sample preparation is highly important, emphasizing that the concentration is the critical parameter. More dilute samples have a low tendency to crystallize; hence, the separation is better in most cases. However, injecting a high volume of a dilute sample causes peak broadening, depending on molecule polarity. If the molecule exhibits a low solubility or high tendency to crystallize, it is recommended to use a more dilute sample or alternatively the "sandwich" injection technique: 50 µL DMSO are loaded right before and after the actual sample into the syringe prior to injecting into the chromatography system, in order to suppress precipitation. Enrichment of the sample with DMSO is also a way to proceed. The "at column dilution" method enables the introduction of poorly soluble or lipophilic compounds to the column. Injection is done by a separate HPLC pump and a valve. The pump transfers the sample dissolved in a solvent with high organic content from 80 to 100% with a low flow rate onto the column. Afterward the valve is switched to the mobile phase with low organic content and higher flow rate to run the actual separation.

6.2 Case Example 1

6.2.1 Investigation of Methohexital Impurities and Decomposition Products

Methohexital is a drug from the class of barbiturates without analgetic effects. It is used as an anesthetic and its formula weight is 262.3 g/mol.

6.2.2 Sample Preparation

To yield a sufficient amount of analyte for NMR analyses by preparative HPLC/MS methods, the sample was subjected to different, though optimized degradation experiments. The optimization of degradation experiments was achieved by continuously adapting the method parameters based on observation values.

Masses related to the main products of the degradation experiment:

$$M + H^+ = 281.2$$
$$M + H^+ = 238.2 \quad \text{(traces)}$$
$$M + H^+ = 237.2$$
$$M + H^+ = 180.2$$

Impurity $M + H = 281.2$

Detailed investigation in three different regions of the TIC (total ion chromatogram) revealed that impurity 281.2 further decomposes to $M + H = 237.2$ and $M + H = 180.2$. One should consider that most impurities undergo further decomposition during handling, chromatographic purification, and workup (neutralization, concentration, freeze-drying).

In most cases only small amounts of the impurities can be obtained. Therefore, it is recommended, after determination of the synthesis and separation parameters (analytical and preparative), to use only a small fraction of the latter for stability testing. Thereby, much work can be spared and the stability of the impurity can be checked under different conditions. In some cases, conditions have to be adjusted to ensure the stability of the impurity, for example, running the separation in a neutral or basic medium. Improvements are not always complicated and should be tested right from the start on the analytical column. Besides the matrix around the target molecule, the effort mostly depends on the complexity of the separation problem. Choosing new parameters might result in co-elution of some present compounds with the desired one. Switching to the preparative column requires some expertise. Even experienced lab personnel might need to draw on personal experience for the given situation Each problem has individual challenges and you need to get "into it" and have a "feeling" for the project. Even in the case of structural analogues, the behavior of compounds might be extremely different.

Sometimes difficult matrices, like PEG, lead to a "smearing" over the column – and this is not dependent on the column type. In these cases, it is recommended to use methanol in the first purification step as the organic eluent, although the separation might be not optimal. Initial conditions should be run with the highest possible amount of methanol that still enables the desired separation. With this step, it is possible to remove 95% of the perturbing PEG. In the second preparative step (after concentrating/lyophilization and dissolution), it is possible to achieve a nearly perfect separation result with acetonitrile as the eluent. High purity solvents like gradient grade are as important as an attentive and accurate operator. A sample (preparative mode) should contain the highest possible amount of water but without risk of precipitation and should be injected with the highest possible concentration (also see above). Analytes with good water solubility should be dissolved in 100% water.

Impurity with M + H = 237.2

This impurity $M + H = 237.2$ further decomposes to the following:

$M + H = 196.15$ and
$M + H = 212.16$

Impurity M + H = 180.2

The impurity $M + H = 180.2$ further decomposes to the following:

$M + H = 196.15$ and
$M + H = 212.16$

The three impurities listed down below were isolated with preparative HPLC/MS methods, subsequently characterized by NMR experiments and, thereby, their proposed structures were verified.

$M + H = 281.2 \, g/mol$
$M + H = 237.2 \, g/mol$
$M + H = 180.2 \, g/mol$

6.3
Case Example 2

6.3.1
Separation of Oligomers from Caprolactam, Multicomponent Separation of Impurities on a Gram Scale

It is also possible to be given the task of obtaining from one sample several impurities in pure form that are present in relatively low concentrations (under 0.5%) and which are mixed with the main product.

In the present case, a caprolactam sample was taken directly from the production process. Six impurities should be isolated in an approximate yield of 1 g each, whereby they accounted for only 0.1–0.4% of the sample.

Following oligomers were identified as the impurities:

Caprolactam dimer	$M + H = 227.1$ g/mol
Caprolactam trimer	$M + H = 340.1$ g/mol
Caprolactam tetramer	$M + H = 453.2$ g/mol
Caprolactam pentamer	$M + H = 566.3$ g/mol
Caprolactam hexamer	$M + H = 679.4$ g/mol
Caprolactam heptamer	$M + H = 792.4$ g/mol

A direct separation of compounds with a high structural analogy can be very challenging because of co-elution. Furthermore, the signals are suppressed by the high product concentration of the caprolactam. A synthesis of the single molecules is economically unprofitable.

In such cases, selective precipitation can be beneficial to deplete the caprolactam. It requires intuition and playing around with solvent systems paying attention to the sample behavior.

After 3 days of extensive testing, a ternary solvent system was found that was able to selectively precipitate the caprolactam but simultaneously kept the six impurities in solution. Thereby, the main component could be completely removed from the solution.

The analytic and preparative method development for the separation of the six impurities was performed after evaporation of the solvent. The analytic method was smoothly transferred to the preparative system, due to the robustness of the oligomers under the developed separation parameters.

6.4
Case Example 3

6.4.1
Preparation and Isolation of bis-Nalbuphine from Nalbuphine

Nalbuphine, (−)-17-(cyclobutylmethyl)-4,5α-epoxymorphinan-3,6α,14-triol, is a semisynthetic opioid with a pain-relieving effect and a molar weight of 377.44 g/mol. It is used in pre- and postoperative treatment in the form of its hydrochloride as an injectable.

During storage, degradation of nalbuphine to its corresponding 2,2′-dimer, bis-nalbuphine, with a molar weight of 712.82 g/mol was observed.

The degradation is the result of an oxidation of the phenol moiety and a subsequent intermolecular condensation to the dimer (bis-nalbuphine). The dimer is sensitive to acidic, neutral, and alkaline conditions. Exposure to light over a longer period of time, as well as temperatures above 40 °C also lead to decomposition. The challenge was the selective synthesis and isolation of 100–200 mg of the desired bis-nalbuphine from nalbuphine as a reference standard. In the first step, a suitable oxidation method that was capable of producing an adequate amount of the desired product for the subsequent chromatographic purification had to be found.

The synthesis was supposed to be trivial but it became a real challenge during method development due to its complexity. Nalbuphine is in equilibrium with the emerging dimer during the reaction and varying reaction conditions to force dimer formation also led to its decomposition.

In cases like this, time is a critical parameter just like the crystallization of some challenging substances. It can take up to 2 weeks to determine the right conditions for the formation of the desired product. Checking on the reaction favored on a daily basis can give hints to find the right parameters. Varying only one parameter in parallel synthesis is a time efficient way to develop the optimal method.

Our choice for the conduction of parallel synthesis are 10 ml pressure resistant microwave vials from Biotage, equipped with security cap and PTFE septum, tempered in alumina blocks (Genevac). These vials are suitable for working under pressure, temperature, or inert atmosphere, taking the necessary safety precautions.

For the analytic reaction controls and method development, the columns Atlantis 3 µm, 100 mm and XSelect CSH 3.5 µm, 100 mm, both from Waters were used. Changing the column led to inversion of retention times of product and starting material. This phenomenon can also appear after pH changes on the same column. It is not possible to predict which column will be the best for each separation problem. It is always the best practice to test different columns to find the most suitable one because compounds tend to behave in ways than you might have expected.

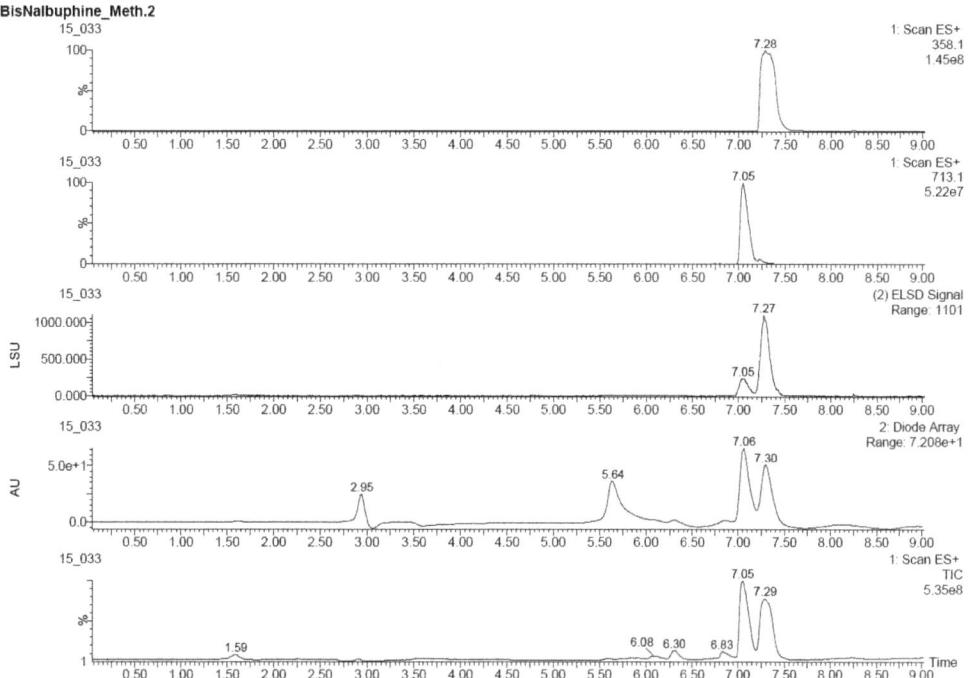

Figure 6.4 Chromatogram of a component after synthesis optimization; comparison between three different detector traces (DAD, ELSD, MS-TIC). Reproduced with kind permission of MCC.

Nalbuphine_Meth.2 is the optimized synthesis method for the oxidation with an incubation time of 1 week at room temperature (Figure 6.4). Another week was required to optimize this method.

In all chromatograms, the ELSD plays an important role in the determination of the ratio between starting material and product (Figure 6.4).

After short workup of the reaction mixture (filtration, cold evaporation, and dissolution in water/methanol), the separation was carried out with MeOH and addition of 0.05% formic acid in water.

The collected fractions were instantly freeze to avoid the degradation until the whole reaction mixture was subjected to purification. The solvent of the combined fractions was removed by freeze-drying.

In most cases, misinterpretations occur concerning the time and resources needed for the production of less known substances (mostly impurities). The most important aspects are the following:

- Stability of the desired substance, for which often no data or literature is available.
- The substance exhibits a completely different stability profile in the analytic system compared to the preparative system. Even after cumbersome optimization of the reaction and analytics, unpleasant surprises can occur while trans-

ferring the method to the preparative system, even when exactly the same parameters are used.

In the case that the conditions have to be altered, other column material might be necessary generating unexpected expense.

6.5
Case Example 4

6.5.1
Isolation and Elucidation of Dopamine Impurities

Dopamine (MW = 153.2 g/mol) is a neurotransmitter from the class of catechol amines and is used among others for the treatment of Parkinson disease. Intravenously, an aqueous solution of the hydrochloride is administered. During the storage of dopamine hydrochloride solution, three impurities in concentrations between 0.2 and 0.5% formed and needed to be isolated, purified, and structurally elucidated (Figure 6.5). In the first step, after optimization of the analysis method, a separation of the three impurities could be achieved on different columns. Three samples of 2 mL each were taken from the original mixture to check the stability at different pH values. The samples were exposed to acidic (pH 3), basic (pH 10), and neutral conditions at 40 °C and subsequent concentration by freeze-drying. The sample was filtered through a 0.45 and 0.2 µm syringe filter and subjected to preparative separation.

As in all cases, the sample should have the highest concentration possible when injected into the system. At the same time, the maximum capacity of the preparative column should be tested for the present case to optimize productivity and to yield as much impurity as possible in the given time.

The masses of the impurities obtained after preparative separation were the following:

1. $M + H = 327.29$ g/mol
2. $M + H = 327.29$ g/mol
3. $M + H = 309.27$ g/mol

Gaining information about the original production process and product formulation as well as the reagents and galenic additives used was very important. Some of the additives present are commonly used for stabilizing the solution or the active ingredient, respectively. Impurities can evolve during storage from the active ingredient through decomposition, catalyzed by additives, storage conditions, and interactions with additives.

In this case, it could be ascertained that the agent dopamine reacts with the additive citric acid resulting in amide formation and yielding the regio isomers (position and constitution isomers, impurities 1 and 2). Both of these can cyclize

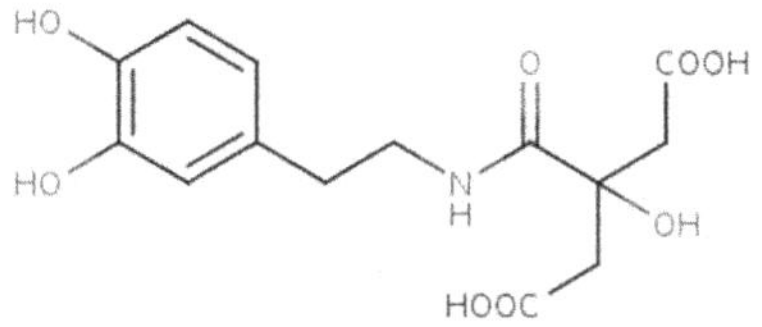

Dopamine Impurity 1; MW=327.2

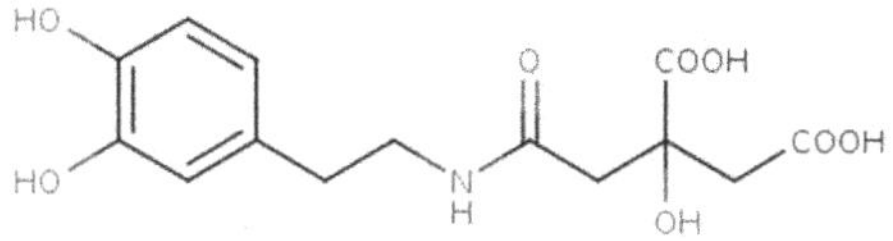

Dopamine Impurity 2; MW=327.2

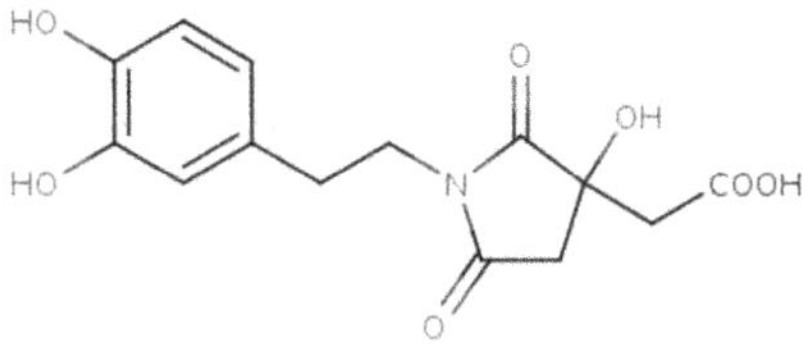

Dopamine Impurity 3; MW=309.2

Figure 6.5 Structures of the dopamine impurities.

to the imide (impurity 3) and react further. The precise structures of the three reaction products were elucidated by NMR experiments.

The information that citric acid was a stabilizer additive of the aqueous dopamine hydrochloride solution supported the proposal of potential structures of the observed masses and fragments thereof.

In NMR experiments, it is recommended to use DMSO-d_6 as the solvent to monitor the exchangeable protons of the impurities. After dilution of the NMR sample with the threefold of water and subsequent freeze-drying, the compound can be recovered in solid form nearly completely. It is self-evident that this method is limited to nonvolatile compounds.

7
LC/MS from the Perspective of a Maintenance Engineer
O. Müller

7.1
Introduction and Historical Summary

This article is about the daily work with a LC mass spectrometer. About 20–25 years ago, GC/MS were mainly used in the routine laboratory. LC/MS was still somewhat new, although already in the 1970s, attempts were made to couple LC with MS [1]. Not until the late 1980s was the first usable device for routine applications available. The difficulty has always been to get ions into the gas phase and to be measured at a pressure of a few millionths of a millibar. If you keep in mind that one mole of water which has a mass of 18g produces a volume of 22.4 L under atmospheric pressure, then you can imagine how this volume would expand under the analyzer vacuum. A small droplet would dramatically increase the pressure during the evaporation in a high vacuum. Theoretically in this case, the transmission of ions would be impossible. LC/MS developers always assumed that the mean free path, which an ion flies in a vacuum, must be sufficient to arrive at the detector without collision. The length of the ion path may be in this case one meter and more. Because of this view, the inlet apertures had been very small with dimensions of less than 0.1 mm. Only later was it found in experiments that ions are also transmitted very well under certain conditions in a higher pressure range [2]. This discovery led to differentially pumped regions and the possibility to use large inlet cross sections. Many improvements over the last instrument generations are based on changes to these assemblies. Formerly, a simple hexapole was sufficient; today a stack of disk-shaped lenses that are controlled by an electronic system are predominantly used. This can add a DC pulse in a range of 0.1–10 V which is jumping with a time offset from disk to disk and moves the ions like a conveyor belt out of the cell. It is also possible to collect the ions at a point and eject them synchronized with the scanning quadrupole. Through these improvements, short dwell times per channel are possible.

The HPLC-MS Handbook for Practitioners, First Edition. Edited by S. Kromidas.
© 2017 WILEY-VCH Verlag GmbH & Co. KGaA. Published 2017 by WILEY-VCH Verlag GmbH & Co. KGaA.

7.2
Spray Techniques

In the early days of the LC/MS technique, the mobile phase was sprayed on a tape or a disc, which was then placed over a heated area where the liquid evaporated. Furthermore, there was a small inlet opening. Today, the evaporation of the liquid takes place in a separate spray chamber which is upstream of the MS entrance. A further development was the thermal spray. The mobile phase is evaporated with hot nitrogen. A significant improvement was the electrospray. Here, a constant electric potential between the capillary and ion inlet is applied. The direction of how the potential is applied varies. With Agilent equipment, the sprayer is grounded and the spray cage is loaded as well as the ion inlet. The opposite is used in Waters devices. Here, the sprayer capillary is loaded and the ion entrance has a comparatively low cone voltage. In the variant of the charged capillary, it sometimes happens that a voltage runs back through the conductivity of the mobile phase. There is a risk that the user get an electric shock. For this reason, a resistor with approximate 30 MΩ is installed in the spray probe. If the charged spray capillary is touched accidentally or by mistake, a majority of the voltage drops down across this resistor. Near the ion source, there is a grounded contact cable which is used to prevent an electric shock. All ion sources have a safety contact switch that disables all the high voltages when somebody tries to open the source housing.

In a spray, the droplets shrink by the evaporation of the solvent with hot nitrogen. However, the number of charges in the droplets remains the same. The system tries to increase the surface area to move the charges further away. From a large droplet, two smaller ones are generated and these are further subdivided. Electrospray is suitable for molecules that are already present as an ion in the mobile phase. When the molecules are nonpolar, the APCI technique can be used. In this case, a constant current flows through a needle. At the tip, a corona discharge takes place, which ionizes gas and solvent molecules. These then transmit their charge to the analyte. All API techniques use cold nitrogen to nebulize the mobile phase. This is important; otherwise, the mobile phase could boil in the capillary. From the outside, a heated nitrogen stream is added to the nebulized spray. It is important to ensure that all gaskets seal well in this area and the capillary exit hole is positioned about 0.5–1.0 mm away from the sprayer tip. Some sources allow adjustment of the sprayer in multiple directions. With the increase of the HPLC flow, the sprayer is further position away from the inlet. If the spray is aimed directly at the inlet, the signal increases but the source becomes dirty in a shorter time.

The mobile phase is sprayed under atmospheric pressure. Overspray exits through openings in the source and is collected in a condensate bottle. The waste gas is connected to a fume hood or vent line. With some SCIEX devices, there is a difference: the excess gas is pumped out of the source by a venturi tube. All devices have in common that only a small portion of the spray enters the MS (only 5–10%). The size of the entrance aperture determines the amount of gas which is aspirated. In a MS with Z-spray source and a sample cone inner diameter (i.d.) of

0.36 mm, approximately 40 L/h were measured at the outlet of the oil pump. In the same time a flow of 800 L/h nitrogen was used for the spray.

7.3
Passage Through the Ion Path

The ions are passed through the instrument by a potential and pressure difference. In the ion block entrance, nitrogen is blown in the opposite direction where the ions enter. This cone gas is used to remove solvent clusters from the molecule ions. The ion inlet always has the same polarity as the ion to be measured and it repels them. This may seem paradoxical, but if it would be the other way around, the ions are attracted to the surfaces and would be lost. Once the ions are drawn, they are bundled by the repulsive effect of the charged walls to a thin beam. This also explains a known effect that is observed in many ion sources. Increasing the voltage at the inlet leads to fragmentation of ions. In this case, they are pushed to the middle of the passage and collide with other molecules like nitrogen. The conditions are similar as in a collision cell. How an ion intake is build varies based on the manufacturers because of patent issues. In general, the MS aspirates the spray through a cone or capillary followed by other conical lenses, so-called skimmer. These separate ions of noncharged molecules. A DC voltage is applied to a skimmer lens. It focusses and passes ions and uncharged molecules are routed to the direction of the backing pump. The path continues through an ion bridge into the next chamber. The ion bridge is usually located in the vicinity of the first turbomolecular pump. Depending on the design, this is a quadrupole, hexapole, octapole, or lens stack. The rods or discs can be operated at a radio frequency (rf) between 0.8 and 2 MHz. The amplitude level increases with the mass to be scanned and is between 50 and 500 V. When ions lose speed because of collisions, a DC voltage is applied to increase the speed. The object of this bridge is to guide ions toward the analyzer; the residual gas is pumped away. After the ion bridge, there is usually an aperture which acts as a restrictor for the vacuum. In Waters devices, this part is built like a cap with a small hole. This cap is grounded with a spring contact. In Quattro Premier instruments, an electrically conductive O-ring is used.

7.4
The Analyzer

In the next step, we look at the analyzer. Here are the quadrupoles, collision cell, and the detector. The ions from the ion bridge fly into the first quadrupole. The driving force was the pressure and potential gradients. Since there is only one pressure range in the analyzer region, no pressure gradient can contribute to the transmission. Similar to the ion bridge, the ions in the quadrupole can be accelerated by a DC voltage. Quadrupoles operate at a radio frequency of 0.8–1.2 MHz.

The amplitude increases in a linear manner during a scan and is approximately 6 kV peak to peak at 2000 m/z. One quadrupole pair is assigned a positive and the other a negative resolving DC voltage. This is about 1/12 of the rf peak-to-peak voltage. All devices have the opportunity to switch this quadrupole polarity via software or electronically. This can be useful if one of the pairs of rods is dirty and a better signal is obtained with the other pair.

Ions can be accelerated if the DC voltage in both polarities is lowered by (0.5–2 V). This parameter is also referred to as ion energy. A quadrupole which runs with pure rf voltage behaves like an ion bridge and directs ions over a wide range. For example, if a quadrupole is set as a filter for a mass of 500 m/z and the resolving DC voltage is set to zero, then the transmission range expands to 80–500%. The quadrupole could pass masses between 400 and 2500 m/z. This property is used in QTOF instruments to influence certain intensities in a spectrum.

The quadrupole is often preceded by a smaller quadrupole so-called prefilter. This has applied a small DC voltage of 5–10 V and can pass through a range of a few hundred Daltons. The majority of all ions are filtered and no longer reach the quadrupole. Thus, premature contamination of the main quadrupole is prevented. In MS/MS devices a collision cell is installed after the first quadrupole. This is a container filled with an inert gas; it has a disk on the front and back side called entrance and exit lens. In the lens there is a small hole with approximate 2–3 mm i.d. through which the ion beam can pass. Inside the cell is an ion bridge. The rods or disks ensure that the ions are deflected from their trajectory by a DC voltage called collision energy. When the applied rf voltage with respect to the collision voltage is greater, then the ions move back toward the center of the cell where they collide with the inert gas and fragment into smaller ions. The gas pressure of about 3×10^{-3} mbar slows the ions down and they can get stuck. A potential gradient between entrance and exit plate accelerates the ions. There are cells in which the rods are arranged conically toward the exit. This is called a linear accelerating cell (SCIEX LINAC). This acceleration works, regardless of whether gas is present in the cell or not. If a spectra is acquired with the second quadrupole, a minimum damping gas pressure is necessary. Otherwise the ions fly simply too fast in the quadrupole and the spectra are poorly resolved.

With the use of columns of small particle sizes, the peaks are becoming narrower and the time which is available for measurement must be shorter. Meanwhile, measuring times of < 10 ms per channel are possible. Older instruments with simple collision cells require at least 50 ms per channel. In practice when a method use with the shortest dwell time of < 5 ms, a small disturbance can already cause deteriorated reproducibility. Often not all analytes are affected in the same way. This makes it very difficult for an engineer to find the exact reason because no error is detected by the electronics. In this case, one workaround could be to increase the dwell time of the affected channels and change the collision gas pressure. Sometimes the results improve when the second quadrupole ion energy is increased slightly. The cause of such problems may be also caused by chromatographic conditions. For this reason, one experiment is to inject a standard mix with a syringe pump direct to the source. If the channels show straight lines in the chromatogram, the MS is functioning and not decreased by ion suppression due to co-eluting peaks.

The detection of ions is carried out by impact on an oppositely charged dynode which is charged up to 10 kV. Electrons are knocked out of the polished metal surface and with some detectors focused by a ring electrode. The detector itself is an electron multiplier or a conversion detector. For the conversion, the electrons are conducted on a positively charged phosphor screen and the resultant flash of light is measured by a photomultiplier.

7.5
Maintenance

We now want to deal with the effects of soiling during operation. In contrast to HPLC, a mass spectrometer is a device that becomes contaminated during operation. The molecules in the spray are deposited on the surfaces. If a very highly concentrated solution is sprayed over an extended period of time, the device is "blind" for small concentrations. In the course of a few hours, the chemical background is slowly lowered again. For this reason, the interfaces are heated. The surfaces clean themselves by the heat. Inorganic salts, for example, phosphate buffers, can only be removed by manual cleaning and should not be used in LC/MS applications. If the soiling becomes too strong, the surfaces can electrostatically charge and deflect the ion beam. This effect is called charging and it is always associated with a signal lowering or a signal loss. When the ESI polarity is reversed and changed back you recognize a signal improvement followed by a steady signal decrease. The dirt is discharged for a moment and charged up again. In this case, only one thing helps: cleaning! Charging can not be predicted on the presence of visible dirt. Sometimes an MS still works perfectly, despite clearly visible soiling. Metallically glossy surfaces can also cause charging. I would like to give two examples.

In the first case, a freshly serviced MS was calibrated with a NaI-CsI solution and then shut down for a week. No signal was present after restarting. After cleaning the sample cone, the full sensitivity was restored.

In the second case, during routine maintenance, a hexapole was placed in a measuring cylinder made of plastic and cleaned with methanol in an ultrasonic bath. After installation, the signal was weaker by a factor of 1000. Presumably, components have dissolved from the plastic and deposited on the rods. Even further cleaning in beakers made out of glass with fresh methanol/acetonitrile did not result in improvements. The bars had to be dismantled and polished with microabrasive paper. The surface was then cleaned with a cotton wool wetted in methanol. It may happen that a MS has a lower sensitivity after maintenance and the signal improves after some time. Components "bake out" in the vacuum, whereby adhering molecules evaporate. It is always asked how to clean components because the performance depends on the machine. The following strategies should be used.

A) Contaminated components made of stainless steel can be mechanically cleaned by using a suspension of fine alumina powder in water and rubbing it with a cotton swab. If no Al_2O_3 powder is available, a stainless steel cleaner such as Stahlfix stainless steel cleaner can also be used. Subsequently wash thoroughly under warm tap water and wash with a dry cotton swab for residue. If the cotton becomes gray/black the wash procedure has to be repeated.

Attention! Gold-plated so-called inert surfaces are very sensitive. Here, mechanical cleaning is only to be carried out very carefully. The gold layer is often only a few micrometers thick and easily rubs off. For this reason, test your cleaning at an uncritical spot where the ion beam does not directly pass. Simply moisten a cotton swab with methanol and press gently.

A further possibility of cleaning these sensitive components is the use of a hot steam cleaner. These devices are available for little money in specialized stores. If you have problems with charging on certain lenses/skimmers, you can easily roughen them with a fine abrasive paper, stainless steel wool, or a glass fiber pencil. The entrance and exit plates of collision cells are pretreated by some manufacturers in this way.

Note: The rods of the hexapoles and collision cells of older Waters devices are made of stainless steel. They can be removed and polished with microabrasive paper (Scotch 3M 12 Micron Lapping Film, www.Schleifartikel.com). When assembling, the rods can be mounted in any order.

B) Cleaning in solvents is carried out after the mechanical precleaning. Components made of stainless steel or PEEK withstand a treatment in 5% aqueous formic acid without damage. If an electronic board which can not be disassembled is attached to the device, a weaker 1% aqueous solution is used. The addition of five drops of detergent concentrate per liter increases the cleaning effect. The advantage of the formic acid is that it has a reducing effect on oxidized metal surfaces and leaves no residue behind.

Other impurities can be easily removed by alkaline cleaners. In practice, Mucasol Universal Cleaner is a tried and tested remedy. This is a cleaning concentrate which is used as a 1% aqueous solution. This goup of cleaners also includes special cleaning fluids such as the Waters MS Cleaning Solution 186006846. After treatment, the components are rinsed thoroughly with distilled water.

C) If an ultrasonic bath is used, make sure that all components are completely immersed in the liquid and hang freely in the vessel. For fastening, you can use old stainless steel capillaries. Particularly sensitive are octapoles from Agilent devices. The fine soldering points could be disconnected by a longer ultrasonic treatment. Instead of the ultrasonic treatment, these parts are bathed in a beaker with Mucasol solution for half an hour and then cleaned with hot steam.

D) In order that the components are free of further impurities, a 1 : 2 mixture of methanol/acetonitrile is finally used. Residues of this solvent mixture can be easily blown off without leaving spots. The dry components are placed on a lint-free cloth or aluminum foil and protected from dust particles. Before installation, the components should be blown off again. Depending on the design and the materials used, parts of the ion path can also be cleaned in acetone or isopropanol.

Figure 7.1 Micromass Quattro Ultima Pt, pollution after the first ion tunnel.

Figure 7.2 Micromass Quattro Ultima Pt, contamination after the second ion tunnel.

Caution is advised when an electronic board is attached or glued to the component. In this case do not use acetone, only isopropanol!

Examples of heavy soiling are shown in Figures 7.1–7.4.

Users always ask how to clean a quadrupole? In general, it is known that this is an extremely sensitive component. The quadrupole rods have a very high surface quality and are precisely aligned. A hair or fingerprint can lead to malfunction. Basically you should handle this component very carefully. It is forbidden to loosen the screws with which the rods are attached or to disassemble the whole compo-

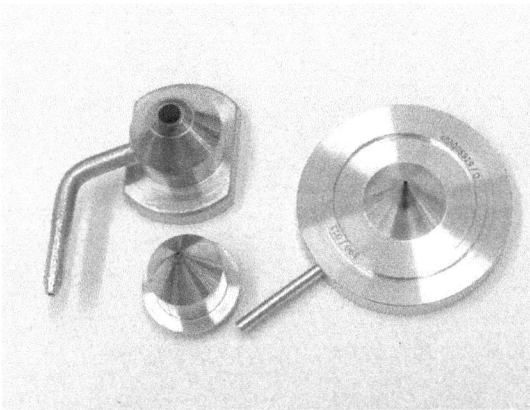

Figure 7.3 Black shadows show contamination of the components of the ion block.

Figure 7.4 Micromass Quattro Premier XE, extremely corroded ion source.

nent. After a few years of operation, dark droplets are visible on a pair of rods. This impurity is called ion burn (Figure 7.5).

These are residues of the ion beam, which can be electrically charged and cause faults. It is often very easy to remove these spots with a cotton swab moistened

Figure 7.5 Agilent MSD, ion burn on a hyperbolic quartz quadrupole.

in methanol. The materials that a quadrupole is made of are different. It may, for example, be made of quartz or ceramic, which is coated with a thin gold film. A stainless steel molybdenum alloy is also widely used. If the residues are burned deeper into the metal, these areas can be polished with diamond paste.

Foam sticks (foamtips) which are also used for the cleaning of electronic components are used as an aid for cleaning the sensitive surfaces. These are moistened easily with solvent. If the quadrupole is very dirty, it may be necessary to immerse it completely in a cleaning liquid. If there are still electronic components such as capacitors or resistors on the quadrupole, these have to be removed first. The quadrupole is then bathed in a warm Mucasol solution for 30 min, washed with distilled water, and then dipped for 15 min in 1 : 2 methanol/acetonitrile. As a rule, a quadrupole usually survives the treatment in an ultrasonic bath. Nevertheless, you should be very careful and consider this treatment only as a last resort. It is possible that coatings will peel off or screws will loosen. The bars should not touch the bottom or side walls of cleaning vessels during all cleaning work. Gloves must be worn during work. The workplace should be clean and tidy. If you respect these basic rules one can dare to clean.

Note: When installing the quadrupole, the connection wires should have a minimum clearance of 10 mm to other components. If the routing of the wires has been altered or the radiofrequency generator exposed to strong vibrations, it may be necessary to check the tuning of the quadrupole RF system. This procedure is also called resonating the RF circuit or dipping the RF. If the oscillating circuit is incorrectly set, the RF generator could switch off by overheating. A symptom would be that from a certain m/z value the spectrum appears to be cut off. In this case the high voltage required for transmission would be insufficient. On Waters devices, the quadrupole operates at a constant frequency. Alignment is used to optimize the resonant circuit consisting of amplifier, coil, and capacitance of the quadrupole. For this purpose, a threaded rod with displaceable metal plates is located in the coil of the RF generator. For Agilent devices, the resonant circuit is tuned via the change of the quadrupole frequency.

After the installation of the cleaned quadrupole, the spectral resolution and signal strength are checked. If the isotope patterns are well resolved, the individual masses are symmetrical, namely, free of "chicken heads or humpbacks". In some cases, the change of the DC polarity results in an additional improvement. If everything is okay, a mass calibration should be carried out.

Cleaning the detector: molecular ions are attracted by the detector dynode. This consists of a cylindrical metal body which is charged with opposite ion polarity. During operation an acceleration voltage between 5 and 10 kV is applied. On impact, electrons are struck out of the metal. The polished surface favors the discharge work. The surface should not be scratched but only wiped off very carefully with methanol-moistened cotton swabs. If the surface has been tarnished by a long service life, the treatment with dilute formic acid is recommended. Electron multiplier tubes which already show reduced sensitivity can be immersed in isopropanol for 15 s and blown off with nitrogen. This procedure is repeated three times. Before switching the multiplier in operate mode, it should dry for at least 2 h in the analyzer vacuum.

Conversion dynode detectors have a different work principle and are largely maintenance free. In some constructions, the dynode is followed by a ring electrode. This lens is used to focus electrons to a beam. Both parts are simply wiped off or immersed in isopropanol. The next component is the phosphor screen. It is very sensitive and consists of a transparent film coated with a phosphorus compound. Electrons flying on this screen cause illumination. This layer must not get wet. This is why it should only be blown very carefully with low nitrogen pressure.

The photomultiplier is responsible for detecting the photons from the phosphor screen and converting them into an electrical signal. This component is usually cast in a glass flask and is maintenance free. The following rule applies: the further a component is removed from the ion inlet, the less the degree of contamination.

In the illustrations above, the sample cone (ion inlet) should be cleaned every 1–2 weeks, the ion block and the following RF lenses annually, and all other components as required (Figures 7.1– 7.4).

Help! Worst case scenario: due to unfortunate circumstances, the HPLC flushed the ion source and the MS sucked in the liquid. The horizontal line in Figure 7.6 indicates the height at which the inlet opening is located. The filling level can be clearly seen from the traces on the lid.

Surprisingly, the ion block and hexapole were only slightly encrusted. The turbopump had not been damaged. The bulk of the liquid was aspirated from the backing pump in the ion block area. As a result, the filling volume of the backing pump increased until it exhausted the excess liquid through the oil mist filter. In the pump, a rust–brown oil/water mixture was present, which had already corroded the cast iron components in the pump interior. After dismantling the oil pump, changing the oil, and cleaning the ion block, the unit was back in working state.

Figure 7.6 Micromass Quattro Micro ion source.

Figure 7.7 Rusty oil–water mixture from the backing pump.

References

1 McFadden, L.C.M.S. (1980) Systems and applications, *J. Chromatogr. Sci.*, **18** 97.

2 Douglas, D.J. and French, J.B. (1990) Mass Spectrometer method and improved ion transmission, patent US4963736A, MDS Health Group Ltd., Canada.

Part IV
Vendor's Reports

8
LC/MS – the Past, Present, and Future

T.L. Sheehan and F. Mandel

Despite the relative high cost of mass spectrometry (MS) instruments, the detection selectivity of MS (ionization process, mass resolution, unique m/z ions from MS or MS/MS processes) and qualitative information of MS (molecular ion, spectral patterns of fragment ions, accuracy mass) has steadily driven the development and adoption of MS technologies by a wide array of analytical applications. MS is the "gold standard" for many laboratories and recognized by international regulatory agencies and courts of law as the source of definitive analytical information. While there have been successful applications of MS with direct sampling, the application of MS has most often been supported by a chromatographic separation by gas chromatography (GC) or liquid chromatography (LC). As far as scope of application for LC and GC, there are many more molecules that are compatible with LC conditions than GC conditions, but coupling LC to MS has faced many more challenges than coupling GC to MS.

GC was fundamentally more compatible with the vacuum requirements of MS. This was especially true as GC column and flow control technology entered the era of high performance capillary columns with the inertness of fused silica and the low bleed of cross-linked bonded phases. Standardization of EI conditions led to libraries of information-rich spectral fingerprints for thousands of compounds, and the advance of computer power allows sophisticated search algorithms to free most analysts from the tedious task of spectra interpretation. By the 1980s, GC/MS was a robust, routine tool providing high quality quantitative and qualitative information for environmental, toxicology, fragrances, food safety, and many industrial labs.

Derivatization chemistries extended the scope of GC applications, but GC/MS can never eliminate the fundamental limitation of analyte volatility in the inlet and column. As a result, many MS labs looked to LC/MS solutions for similar performance, simplicity, and ruggedness. Numerous research groups took on the challenge of interfacing LC to MS with configurations such as direct liquid introduction, corona-discharge source, moving-belt interface, particle beam, and thermospray, but none of these techniques were able to fulfill the expectations for high quality, robust operation. The introduction of the first commercial APCI in 1989 started a revolution for LC/MS, but it was not until commercial electrospray

The HPLC-MS Handbook for Practitioners, First Edition. Edited by S. Kromidas.
© 2017 WILEY-VCH Verlag GmbH & Co. KGaA. Published 2017 by WILEY-VCH Verlag GmbH & Co. KGaA.

(ESI) entered the market the next year that the growth rate for LC/MS began to soar.

Both of these atmospheric pressure ionization (API) techniques have limitations. Ion–molecule APCI reactions only apply to some compounds. ESI worked well for a wider scope of compounds and extended the effective mass range of the MS with multiply charged ions for macromolecules; however, many LC separations (reserved phase, ion exchange, size exclusion) were incompatible with ESI since the sophisticated HPLC ternary and quaternary gradients, array of buffers, and deactivating agents like nonyl amine suppressed ESI response. Even with these limitations, API devices made LC/MS a robust reality and, in turn, accelerated the development of complementary technologies for LC (1 mm, capillary and nano-LC) and MS (quadrupole MS/MS, TOF, Q-TOF, and miscellaneous ion traps). In terms of the principles of "crossing the chasm", LC/MS has experienced a tornado of development with some LC/MS systems entering the "main street" of routine analysis.

With all of this success, now is definitely not the time to take a break from technological developments. In fact, the excitement and enthusiasm surrounding LC/MS demands even more attention to the science and engineering of mass spectrometry. The "low hanging fruit" may have been harvested, but there are still many opportunities for expanding or extending the scope of LC/MS applications and altering the productivity of the LC/MS lab.

Productivity or workflow optimization has become a common point of concern for many labs. Twenty-five years ago, an LC/MS discussion might have asked whether the MS sensitivity was adequate to meet method detection limits (MDL). Today, MS sensitivity often supports detection far below the MDL. As a result, the "excess" sensitivity of modern MS can be applied to drive other benefits such as reduced cost by collecting and preparing smaller samples or extended column life and reduced maintenance by injecting less sample on-column. In turn, this opens opportunities to revolutionize on-line, automated sample preparation. The pursuit of increased sensitivity will definitely continue, but now the pursuit will consider a holistic view of the workflow and productivity.

Cost is another point of concern: both purchase price and cost of operation. In terms of price versus performance, LC/MS is actually much cheaper today than even 10 years ago – especially true when cost is adjusted for inflation. But LC/MS systems are not inexpensive. MS systems have not achieved the reliability and price reduction of commercial electronics such as flat-screen television and mobile phones, but Agilent recognizes the necessity of decreasing cost of maintenance, decreasing the frequency of component failure, and lowering the purchase price to fit into more budgets. Agilent believes MS performance and capabilities will increase, but cost of ownership will trend downward.

While sensitivity is often considered to be the prime performance parameter for MS, detection selectivity is always a close second. For the analysis of complex matrices, unit mass resolution MS (single quadrupole) is being replaced by more selective MS/MS or high-resolution MS systems. Increased selectivity increases confidence in the quality of the quantitative and qualitative information, and increased selectivity also supports changes to workflow similar to sensitivity. If applied correctly, selectivity can save time (faster separations) and reduce operational costs (less sample cleanup before injection). As technology continues to evolve, MS/MS will reach a plateau. The more important evolution may prove to be the introduction of lower cost, high resolution MS and MS/MS. If these high resolution systems also deliver the qualitative benefits of accurate mass, the future of high resolution MS could change dramatically. Evolution could also carve out a position for technologies like ion mobility MS (IMS). Although IMS is only in the hands of early adopters now, the new dimension of information generated by IMS is another technology that changes the selectivity of the MS measurement and delivers unique details not available with non-IMS systems.

In any discussion about the future of LC/MS, one should never forget the LC. Reduction in the mobile phase flow rate with narrow ID columns has definitely increased the robustness of the MS interface, and ultralow nano-LC has demonstrated remarkable increases in sensitivity for sample limited scenarios. Although 2D-LC has had limited success in the past, advances in LC hardware design and instrument control are creating realistic opportunities for routine 2D-LC/MS. For LC-ESI-MS, 2D-LC/MS provides more than the added selectivity of two separations; the second dimension can provide a protective isolation of ESI-incompatible mobile phase due to detection of compounds of interest. Analysts will be able to exploit the benefits of new LC stationary phases tailored for macromolecules without concerns about loss of ESI response.

With all of this attention to hardware, one must not forget the power of computers and software. Where do most analysts spend the majority of their time? At the PC workstation. How can analysts mine information from complex sets of data? With better MS software. Software has already changed the world of mass spectrometry, and Agilent realizes that trend will absolutely continue. Deconvolution, principle component analysis, integration of MS data with biological pathways, compounds extraction from high-resolution data sets and other computations will become as automatic and routine as simple peak area processing or library searches. Result dependent logic will secure workflows without operator intervention. Very large files from high-resolution analyses will be compressed while conserving the information of value. Isobaric interferences and response saturation will be handled by automatically changing to other ions without these problems. The net result will be increased use of MS for pharma, biopharma, clinical, toxicology, food safety, environmental and other application and more time for laboratory personnel to address tasks that still require human intervention – and through it all with higher and higher quality data.

9
Vendor's Report – SCIEX
D. Schleuder

SCIEX (**SCI**entific **EX**change) was founded in the 1970s by three Canadian scientists in Toronto with the aim to develop highly specialized instruments which were later used (among others) in the American NASA Viking probe for Mars exploration. In 1981, the first mass spectrometer (TAGA 6000) was used in a mobile bus to detect pollutants in the air quantitatively while driving. About 8 years passed before SCIEX (called PE SCIEX at that time) introduced its first commercially available mass spectrometer (MS) at Pittcon, the world's leading trade fair conference for laboratory equipment. At this time it was strongly believed by a recent market survey that the world's estimated need for mass spectrometers is not more than 100 mass spectrometers worldwide—as of today, there are now tens of thousands with growing potential.

Mass spectrometry is unconceivable without a coupling to liquid chromatography (LC). Both technologies have evolved in parallel in recent years and complement each other in a way that new workflows – especially adapted to the increasing market demands – can be realized. Today ionization techniques like electrospray ionization (ESI) or chemical ionization at atmospheric pressure (APCI) are widely used. The LC flow rates normally vary between about 200 µL/min and 2 ml/min for high flow applications, whereas nano-LC/MS systems operate with flow rates between 200 and 1000 nl/min. Analytical labs like food, pharmaceutical, environmental, and forensic labs generally use their LC/MS instruments in high-flow mode routinely. However, the excess of LC solvents must be removed before ions enter the mass spectrometer. In previous SCIEX instruments, the LC flow was introduced almost axially and the solvent was dried by a single heater. In today's SCIEX devices, this happens orthogonally and two heaters are used (Turbo V™ ion source). Moreover, the patented SCIEX "curtain gas" (nitrogen gas flowing against the ions) desolvates the aerosol through collisions with the residual gas molecules without ion fragmentation. The curtain gas technology is also used in all modern SCIEX instruments.

In mass spectrometry, triple quadrupole (QQQ) MS/MS systems are often used. Their strength lies in the quantification of, for example, pesticides, mycotoxins, drugs, steroids, pharmaceuticals, and many other substances in the trace range.

The HPLC-MS Handbook for Practitioners, First Edition. Edited by S. Kromidas.
© 2017 WILEY-VCH Verlag GmbH & Co. KGaA. Published 2017 by WILEY-VCH Verlag GmbH & Co. KGaA.

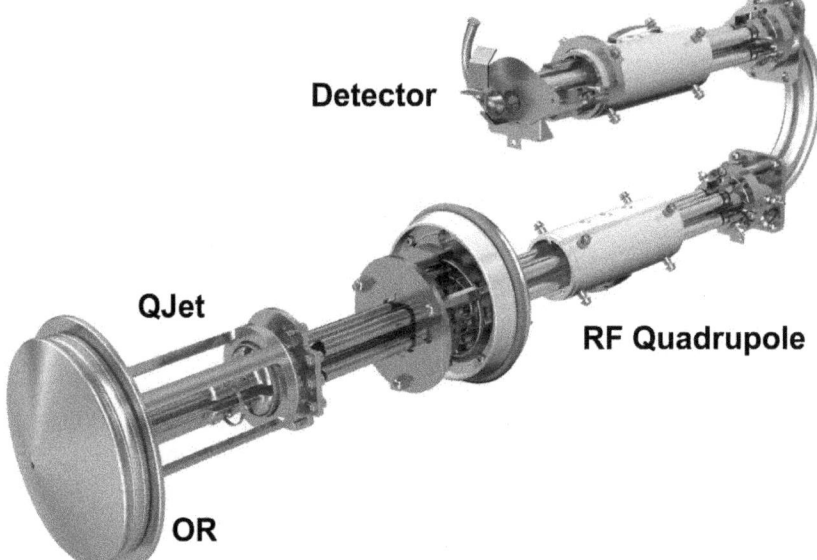

Figure 9.1 QTRAP® 6500 System ion path: curved collision cell with orifice (OR), QJet (Q0) for ion transfer, RF quadrupoles, and detector. Reproduced with kind permission of SCIEX.

The matrices range from food and drinking water over packaging materials to blood plasma, serum, or urine. The analysis is usually carried out in the so-called MRM mode (multiple reaction monitoring). After LC separation and ionization of the substance, the precursor ion will be selected in the first quadrupole MS, the generation of the fragments take place in the second quadrupole (collision cell), and the third quadrupole selects fragments from the precursor ion. For this mode, the knowledge of the mass-to-charge ratio of the precursor ion and the fragment mass is required (targeted analysis). This is currently the most sensitive and fastest LC-MS/MS technique for the quantification of ionized substances with excellent reproducibility and large linear dynamic range. Such a QQQ system with a curved collision cell is displayed in Figure 9.1 as an example.

With the development of small LC column particles (sub-µm) and the trend toward even shorter LC run times, the demands on the LC-MS/MS systems have increased. For the MRM mode, new algorithms have been developed. With these, the user has the option to couple the MRM transitions of an analyte with its retention time. At SCIEX, this algorithm is called "scheduled" MRM (sMRM™). Latest software developments even allow dynamic adjustments to compensate for any instability of the retention times. Besides the possibility to record several hundred parallel MRM traces, this also enables the possibility to shorten LC run times without compromising the number of data points necessary for accurate quantification (Figure 9.2). Usually two MRM traces are used for the analysis of a substance: the peak area of the so-called quantifier is used to quantify the substance. The signal intensity of the second MRM trace (qualifier) must have a

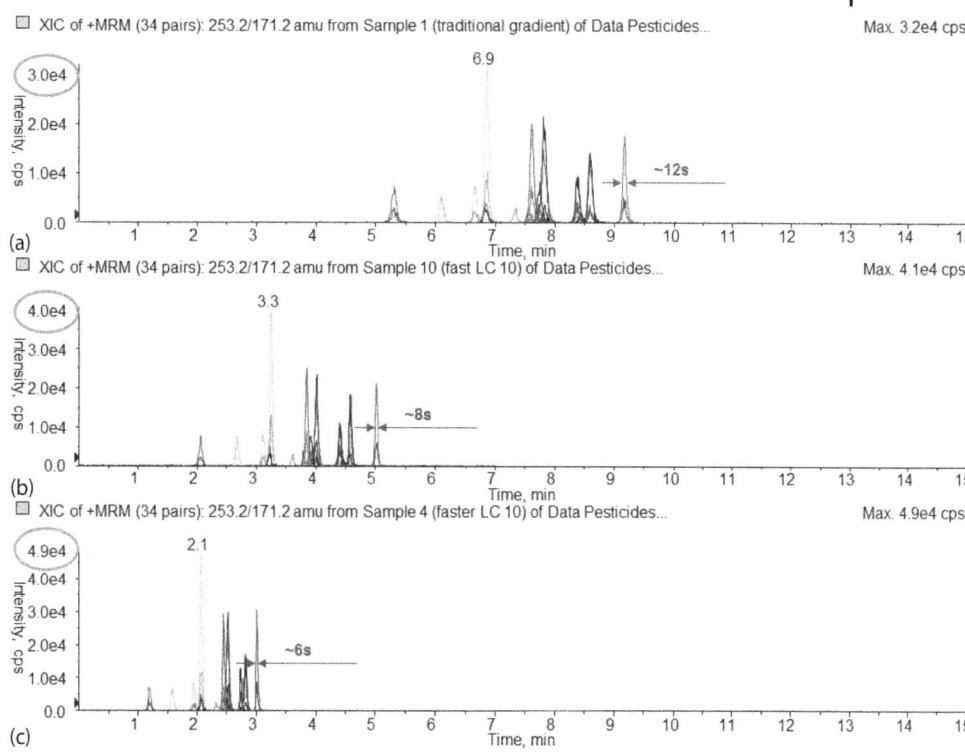

Figure 9.2 Comparison of "traditional" (a), fast (b), and UHPLC gradients (c). Reproduced with kind permission of SCIEX.

specific ratio to the intensity of the quantifier signal. Together with the retention time the confirmation of a particular substance is unique.

A further development of the QQQ systems is capable to acquire MS/MS besides MRM traces with the so-called "full-scan mode." With classical QQQ instruments, this approach is very limited. At SCIEX, these instruments are called QTRAP® systems (linear ion trap) and are defined as hybrid systems. In IDA experiments (information dependent acquisition), data can be recorded to quantify and identify substances in parallel in a single LC-MS/MS run. Either a MRM run or a full scan run triggers the MS/MS spectrum. The MS/MS spectrum can then be used to identify the substance with the help of spectral libraries. This is a helpful approach whenever a substance, for example, does not have a proper qualifier or the intensities of the qualifiers are not sufficient.

The QTRAP® technology also permits the quantitation via MS/MS/MS experiments (MRM³), which occurs as another fragmentation step in the third quadrupole. As for the classic MRM mode, a precursor ion is selected in the first quadrupole and will be fragmented in the collision cell. After that a particular fragment is selectively enriched in the third quadrupole (which serves as a linear ion trap) before it is fragmented via resonant excitation. The detected MS^3

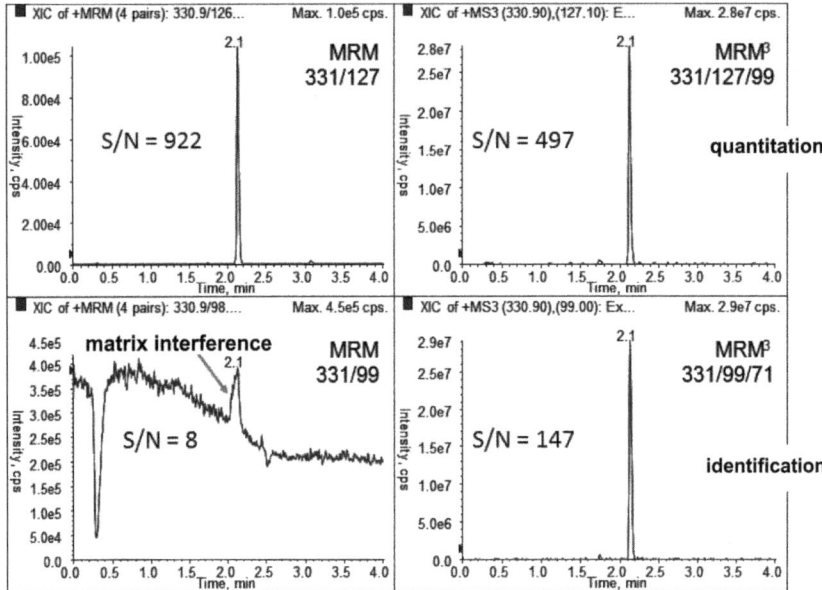

Figure 9.3 MRM transition 331/127 can be taken for quantitation, while the MRM transition 331/99 shows matrix interference. MRM^3 transition 331/99/71 has almost no matrix interference. Reproduced with kind permission of SCIEX.

fragments may be used for highly selective quantitation or for structural analysis. As an example, one MRM and the corresponding MRM^3 trace is displayed in Figure 9.3.

Recent developments in the LC-MS/MS technology such as other hybrid systems with QTOF architecture (TripleTOF® and X500R QTOF) are now offering the possibility of a qualitative and quantitative screening in routine. This hybrid technology combines the advantages of QQQ systems with those of the high-resolution "time-of-flight" (TOF) analyzers in a single mass spectrometer. In addition to very good sensitivity, this also allows a high mass accuracy and a large linear dynamic range. These systems also operate in both the TOF-MS mode and TOF-MS/MS mode. In TOF-MS mode, the precursor ion is transferred directly without fragmentation into the TOF analyzer and its mass is accurately determined based on its flight time. However, in MS/MS mode the precursor ion is selected in the first quadrupole and the precursor is then fragmented in the second quadrupole. The fragments will be transferred into the TOF analyzer and the mass-to-charge ratio (m/z) can be accurately determined according to the flight time. As for the QTRAP® system also here a parallel detection of MS and MS/MS data in IDA mode takes place. The resulting high-resolution datasets can be analyzed following two different strategies: targeted and nontargeted. For the targeted analysis a mass extraction window of 5–20 mDa will be set to quantify the analytes of interest with high selectivity. The exact MS mass and the exact MS/MS masses allow a confident identification of the target analytes. Based on the MRM mode

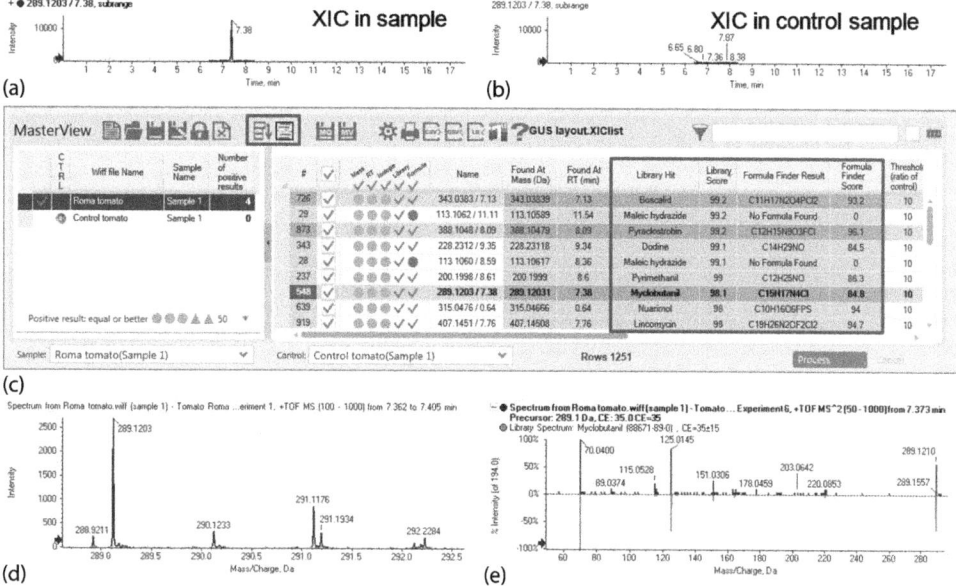

Figure 9.4 Extracted ion chromatogram (XIC) from a sample (a) compared with an XIC from a control standard (b). Identification with exact mass, retention time, isotopic pattern and the calculated sum formula (c). MS spectrum (d) and MS/MS spectrum (e) in comparison to a MS/MS spectrum from the spectral library. Reproduced with kind permission of SCIEX.

for QQQ instruments SCIEX has named this scan as MRM^{HR}. For the nontargeted analysis the exact mass, the isotopic pattern and potential adduct information can be used to calculate the empirical sum formula. The exact MS/MS masses can furthermore be used for the identification of the substances or to determine the fragmentation path. This is shown in Figure 9.4 as an example.

An extraordinary advantage of a X500R or TripleTOF® instrument over the QTRAP® technology is the presence of full and highly resolved MS and MS/MS data over the entire chromatogram. These data can also be used for retrospective analysis. Especially in the food, environmental (e.g., water supplier), or in forensic domain, a complete MS and MS/MS dataset at any retention time will be provided to the operator. Unexpected, organic substances in trace ranges can be tracked statistically by comparing samples deriving from different times (kinetic study) or abstraction places. On the basis of MS and MS/MS data, a substance can be identified via spectral libraries and quantified whenever a reference substance is available.

Both the QTRAP® and the TripleTOF® technology can be equipped with a differential ion mobility spectrometry cell. SCIEX calls this technology the SelexION® technology and allows the separation of ions via its ion mobility under atmospheric pressure. The advantage of these cells can be seen in the separation of isobaric compounds, substances that co-elute from LC, or analytes that suffer from high background.

In these two approaches, both the SelexIon® technology and the QTOF technologies (TTOF & X500R), SCIEX sees the biggest potential for the future with growing market demands.

10
Manufacturer Report – Thermo Fisher Scientific
M.M. Martin

In recent years, the world of separation sciences has seen a lot of changes. The introduction of UHPLC has reboosted liquid chromatography remarkably at a point in time where HPLC was seen as a mature and somewhat stagnant technology; the same applies to mass spectrometry, where Orbitrap technology as a new kid on the block has reset the mindset of many LC/MS users for more than 10 years now. However, joint LC/MS systems still require solid expert knowledge and understanding of the underlying processes, particularly if the user needs to critically evaluate the results of such an analysis. For the next future, we expect a differentiation of the hardware offers into two directions: On one side we will continue to see professional expert systems combining maximum performance with a high degree of flexibility for a tailored configuration matching best the individual customer application requirements. They will be the ideal tools for high-end research and development work. On the other side, LC/MS technologies will progress to expand into fields of work where the users typically do not have – and should not need to have – expert skills in both LC and MS technology; this will drive the development of integrated analysis systems for strictly predefined applications or workflows, but with a substantially simplified user concept, allowing also nonexperts to successfully generate meaningful results with only little training effort. Nobody who wants to get a driver's license today needs to understand in detail how a combustion engine works – it should be the same for the use of these LC/MS analyzers; only this simplification (also) for nonexpert users will help LC/MS technologies to expand into new markets. In addition to the traditional pharmaceutical and biopharma industries, which have been LC/MS strongholds from the early days of this technology, medical applications in personalized medicine, therapeutic drug monitoring (TDM), and clinical diagnostics will be the main drivers for growth. However, new instrumentation first needs to overcome the ambitious hurdles of medical device registration before entering these attractive and prospering markets.

The HPLC-MS Handbook for Practitioners, First Edition. Edited by S. Kromidas.
© 2017 WILEY-VCH Verlag GmbH & Co. KGaA. Published 2017 by WILEY-VCH Verlag GmbH & Co. KGaA.

10.1
Liquid Chromatography for LC/MS

UHPLC is meanwhile well accepted as a worthy extension, even a successor of conventional HPLC; and for good reasons, UHPLC systems are established as the new frontend standard in LC/MS applications in particular. First and foremost, LC/MS analyses strongly benefit from the extended capabilities of UHPLC: enhanced LC resolution improves the quality of MS results, while high-speed separations allow for a better workload and less idle times for cost-intensive, high-value mass spectrometers; and the extra charge for a UHPLC system compared to a HPLC device is very moderate compared to the substantially higher investment in the mass spectrometer. Many years of intensive research and development on the UHPLC instrumentation side have resulted in substantially improved reliability of UHPLC equipment and robustness of results, thus, significantly lowering downtime of LC/MS installations where the mass spectrometer is forced to inactivity every time the frontend is malfunctioning. The technical challenges of handling the elevated pressure and, thus, the energy input into the separation system have led to many creative technical innovations which in summary took the precision of UHPLC separations to the next level. Retention time precision significantly lower than 0.01% is easily achieved even at maximum pressure load of up to 1500 bar/22 000 psi in 24/7 routine analysis; sophisticated injection mechanisms like the intelligent *SmartInject* technology of the Thermo Scientific™ Vanquish™ UHPLC systems virtually eliminate the pressure shock during the sample injection, thus, enabling superb precision levels and extending the UHPLC column lifetime at the same time. This easily translates into UHPLC columns lasting at least as long as conventional HPLC columns operating at moderate pressure in the 200 bar regime. Modern fitting systems like the Thermo Scientific Viper™ fingertight fitting technology allow for a tool-free, finger-tight installation of both system tubing and LC column, repeatedly and reliably even at pressures up to 1500 bar/22 000 psi, while practically excluding any additional unwanted void volumes in the fluidic connection. This also avoids unexpected leakages without applying brute force with unwieldy wrenches, and the high-performance separation of a state-of-the-art UHPLC column is seamlessly translated into a high-quality LC/MS chromatogram with efficiencies of more than 300 000 plates per meter column length. Due to the high pressure reserves of modern UHPLC instrumentation, for example, up to 1500 bar/22 000 psi for the Vanquish systems, this famous "meter column length" is no longer a vague theory, but it becomes a practical reality for ultrahigh resolving chromatography by creating daisy chains of four or more 250 mm UHPLC columns. Highly optimized optical detectors based on fiber optics like the Thermo Scientific LightPipe™ technology make trace level analysis accessible for UV detection, a domain where mass spectrometry has been essential for sensitivity reasons in the past.

In addition, mass spectrometry will certainly continue to evolve as the new detection standard also in routine analysis. In this context, LC separation columns of 1 mm inner diameter (I.D.) can be a useful tool to bring the demand for chromatography running at ESI-compatible flow rates in the lower µL/min range in line with the performance of sub-2-µm particle columns which ask for mobile phase linear velocities of 5–7 mm/s and beyond to stay in or exceed the van Deemter optimum. However, packing these columns is a critical process which explains why achieving both good batch-to-batch reproducibility and long-term stability of these columns is still a challenge today. Given a reliable manufacturing procedure for these 1 mm I.D. columns, they can be quite attractive for LC/MS analyses. A progress on the column side will then need to come along with significantly reduced gradient delay volumes (GDV) on the instrument side. Current UHPLC systems typically still have a minimum system GDV of approximately 100 µL, which remarkably limits the speed-up capabilities at flow rates lower than 100 µL/min. At the same time, it means that it is extremely challenging to use the same pumping technology to serve a flow range from a few dozens of nanoliters per minute (for nano-LC) up to a few milliliters per minute (for wide-bore LC columns), free of pulsation, at pressures beyond 1000 bar. Therefore, the universal UHPLC flow delivery, which means one single pump device which drives every LC separation from nano-LC/MS to analytical scale UHPLC-MS, will hardly be realized in the foreseeable future. Thus, it is expected that the LC/MS market will continue to see two different UHPLC system philosophies: on the lower end of the flow range, nano-/cap-LC systems which dominate the fields of proteomics and metabolomics in combination with MS detection will most likely extend their upper flow limit to better operate also with 1 mm I.D. columns, which have a higher loadability than classical cap-LC columns. On the other side, analytical scale UHPLC systems will slightly expand their lower flow rate limits so that users of classical 2.1–4.6 mm ID columns for small and large molecule analyses in pharma, biopharma, environmental, and food safety applications will also have the chance to profit from the benefits of 1 mm I.D. columns.

10.2
Mass Spectrometry for LC/MS

As stated before, mass spectrometry will certainly be further established as the new detection standard in many markets. However, we will most likely not see a new, disruptive mass spectrometric technology in the near future. The latest introduction of an entirely new mass analyzer design, Thermo Scientific's Orbitrap™, has celebrated its 10th anniversary. Since the Orbitrap launch, this MS type has redefined the entire market for high-resolution/accurate mass (HRAM) measurements, and it has placed a lot of innovation pressure on the older competing techniques like time-of-flight mass spectrometry (TOFMS). The increasing market share of the Orbitrap technology, on the one hand, and the rising data quality demand in clinical diagnostics, customer healthcare, environmental and consumer protection, on the other hand, also pushed HRAM applications into routine segments in recent years. This trend will certainly grow further, as HRAM datasets provide a much higher level of information detail which translates into higher confidence for compound identification and lower probabilities for false-positive results. Hence, high-resolution mass spectrometry will raise pressure on established technologies like quadrupole MS (primarily triple quad MS). We expect the Orbitrap technology to be a key player here due to their inherent advantages compared to other high-resolving mass analyzers. It has both higher robustness and superior performance than TOF technology and scales up easily for higher resolving power. TOF analyzers struggle with improvements on both mass accuracy and mass resolution; even with extensively long drift tubes or sophisticated ion reflectrons, they barely reach a performance level which is only at the lower end of what Orbitraps can realize. Resolving power at a level of ion cyclotron resonance MS (FTICR) is by far not achievable with the TOF techniques of today, but is definitely in the range of what Orbitraps can do; but in contrast, Orbitraps do not need the elaborate and expensive infrastructure and the high investment required to operate FTICR-MS. We already discussed earlier that the mass accuracy of TOF devices is strongly affected by drifts and fluctuations in the ambient temperature, which can only be counteracted effectively by a substantially higher calibration effort (Chapter 2). TOF devices still offer some advantage with respect to data collection rates, scan speeds, and price, but this lead is still quite small.

As a reaction to the expansion of HRAM mass spectrometers breaking into the domain of triple quad MS devices, the triple quads will develop away from pure expert systems and more into the direction of sophisticated LC detectors. Their main field of work in the future will be the simple, cost-effective, and robust quantification of analytes in samples of moderate complexity. This, however, requires a significantly facilitated user interaction and guidance also for less experienced operators. The current performance level of triple quads is already more than sufficient for the vast majority of applications so that further improvements of the technology do not automatically translate into a higher customer value. Modern triple quad devices offer, for instance, a sensitivity enabling limits of detection which are orders of magnitudes lower than required by current regulatory authorities, clinical trials, or pharmacological activity levels. Hence, lowering the detection limits even further would not really pay off for the user, while higher scan speeds or shorter duty cycles would, especially with respect to UHPLC separations. HRAM experiments, however, are beyond the scope of quadrupole technology anyway due to inherent limitations of the operation principle. Therefore, one future trend will certainly consist of more compact and robust quadrupole analyzers, featuring a high degree of integration into the UHPLC system, a massively simplified user interface, and following a virtual "no-maintenance" philosophy, at a cost of some limitations in performance and flexibility.

10.3
Integrated LC/MS Solutions

It is an open secret that the current state of integrating LC, MS, software, and consumables into one single solution has some room for improvement, across the board of all manufacturers and solution providers. Primarily driven by contract research and manufacturing organizations (CROs, CMOs) as well as quality control (QC) labs, a very prominent trend in recent years asked for LC/MS instrumentation which no longer requires expert knowledge to be operated successfully. As a consequence, there is an increasing offer of total system solutions or "analyzers" which combine selected LC and MS hardware with optimized and validated separation methods, predefined LC columns, reagents and solvents, and tailored software solutions including readily implemented electronic workflows into a complete package for dedicated analytical workflows. The set of ready-to-use consumables for sample preparation, tailored LC columns, sample containers, reagents, and eluents, thus, strongly improve the method robustness and the quality of results and reduce the negative impact of user errors, the user workload, and the cost per sample. Therefore, the market will surely see further increasing offers of such complete workflow solutions. Depending on the application field, one could even think of compact, well-integrated instrument designs, going beyond what is already achievable today, with the Thermo Scientific EasySpray™ nano-LC interface as a prominent example. All this highly integrated and tailored instrumentation, however, will then have a very limited range of applicability. Hence, the "analyzers" will be perfectly suitable for routine users with a clearly defined field of work and application; they certainly will not be flexible enough for high-end research and development work, where freely configurable and modular LC/MS expert systems will remain at the forefront.

10.4
Software

If one thing is for sure, then it is that software will play a pivotal role for all future LC/MS technologies, as it forms the central user–machine interface; in addition to a nice and appealing control surface, reducing the complexity of operating sophisticated (U)HPLC-MS instrumentation will be of fundamental importance. Currently, two opposing software perspectives dominate the global market for LC/MS data acquisition and analysis.

First, there are the chromatography data systems (CDS), which were originally written for the control and programming of (liquid and gas) chromatographs as well as for the acquisition and analysis of chromatographic data. Over the years, many of these CDS have evolved into highly professional data management systems offering multivendor instrument support and operating in global network environments (wide area networks, WAN). They fulfill all state-of-the-art regulatory requirements for compliance, traceability, and data integrity. However, while being experts in handling chromatography data, the integration of mass spectrometry control and data analysis has not been in the focus in the past. In the worst case, users were required to work with two software packages in parallel, one controlling the UHPLC system, the other the MS instrument, using hardware synchronization via a signal trigger cable, and including doubling everything, from instrument methods, sequences, and data files to the efforts for data analysis and archiving.

Second, all notable MS vendors offer dedicated MS instrument control and data analysis software, which, however, typically downplay the control capabilities and settings of the LC frontend. As a consequence, LC instrument parameter tracking as well as data analysis tools reporting chromatographic performance data are mostly implemented with major limitations. This might not be perceived as a serious deficiency in the everyday routine work, but when it comes to troubleshooting support, this low-level implementation quickly reaches its limits.

In recent years, the leading CDS have made great progress with the integration of MS instrument control and data analysis. In particular the two market leaders, Chromeleon™ 7.2 from Thermo Scientific and Empower™ from Waters, offer a widely extended MS support. Chromeleon, for instance, supports both GCMS as well as LC/MS data acquisition and evaluation, with a broad selection of triple quadrupole and Orbitrap instruments on the LC/MS side. Another key strength of Chromeleon is its full compliance with all international regulatory requirements. GxP compliance is in general a feature which will see a rising demand, not only in the pharmaceutical industry where compliance had its origin, but also more and more in other fields of consumer protection like environmental and food safety analyses.

It is very likely that in the future, the boundaries between LC, GC, and MS data handling will become increasingly blurred. Especially the GCMS integration has been a great role model in the past, but LC/MS and ICMS integration are in the meantime rapidly catching up. It is to expect that this ongoing trend will finally result in a unified "C/MS data system" which hosts really all kinds of chromatographic and mass spectrometric data including full instrument control support. At the same time, these software solutions will offer a higher degree of integrated intelligence for easier operation, simplified method transfer, and reduced instrument downtime by smart maintenance prediction. Particularly the maintenance and support scheduling for a maximum degree of instrument availability will strongly benefit from global network integration. Remote device monitoring helps the early identification of deviations in instrument parameters from the normal and, thus, to prevent unexpected device malfunctions and longer downtime. In the case of a hardware failure, the remote functions enable service engineers to identify the root cause of the error prior to the onsite visit, which reduces repair time and service costs remarkably. Depending on the corporate IT security requirements, the omnipresent trend of cloud computing in IT applications will also lead to new capabilities in analytical sciences. Cloud systems for MS data evaluation and database search of mass spectra are already available today (like mzcloud.org, supported by Thermo Fisher Scientific [1]), as it would only be the next logical step within an industry-wide trend to implement more artificial intelligence into the cloud, for instance, to automate and facilitate complete LC/MS workflows. Cloud storage of instrument methods or workflows can also be very helpful for corporations to deploy standardized analytical procedures globally and make them accessible worldwide. The AppsLab™ library from Thermo Scientific [1] gives some sneak preview today of what could be realized with an efficient network infrastructure and a smart user interface. Simultaneously, the global network capabilities for distributed work will certainly reinvigorate discussions dealing with the aspects of IT security and minimum requirements to protect sensitive company data and business secrets. Company-internal cloud solutions, protected from external access, can be a viable and attractive option here. Overall, users will be able to benefit from a large number of different offers which decouple the analytical task from the lab where the instruments run. Net-based system run control and analysis of C/MS data around the globe, for instance, with users in Europe having immediate and virtually real-time access to data generated at a site in Asia, will be the new norm, along with device control and data visualization on portable devices like tablets and smartphones. Together with the previously mentioned remote diagnostics, maintenance features, and electronic lab management, analytical services will disconnect more and more from the direct contact with the real instrumentation and continuously evolve into a virtualized lab. A much better interfacing of the C/MS systems with lab information management systems (LIMS) will finally help to gain a comprehensive understanding of the identity of chemical compounds across many analytical techniques and company sites on the planet, enable seamless data traceability, and thus ultimately take the quality of analytical information to the next level.

References

1 HighChem LLC (2013) *m/z* Cloud Advanced Mass Spectral Database, www.mzcloud.org (accessed 10 March 2017).

2 Thermo Fisher Scientific Inc. (2017) Thermo Scientific AppsLab Library, https://appslab.thermoscientific.com (accessed 10 March 2017).

About the Authors

Claudia vom Eyser
Claudia vom Eyser finished her studies of water science at the University Duisburg–Essen in 2011. After graduation, she received her PhD at the faculty of chemistry "Instrumental Analytical Chemistry" at the University Duisburg–Essen. Since 2011, she works as research associate at the Department of Environmental Hygiene & Micropollutants at the Institut für Energie- und Umwelttechnik e. V. (Institute of Energy and Environmental Technology). Her focus of research includes the analysis of complex environmental samples like sewage sludge and biochar using liquid chromatography mass spectrometry. In addition, the development of sample preparation techniques and its automation are of great interest.

Edmond Fleischer
Chemistry engineering studies at the Fresenius College Idstein. PhD candidate at the University of Mainz, Department for Chemistry and Pharmacy.
Subject of graduation: Structure–activity relationship of synthetic and natural compounds in their anti-tumor activity. After 14 years of industrial experience in the pharmaceutical research, Edmond Fleischer took over the laboratory management at Microcombichem MCC in Wiesbaden.
Among his main activities for the last 5 years belong the tailored synthesis with pharmaceutic character, the purification and structure elucidation of active agent impurities, innovative research projects in cooperation with universities and formulation of cosmetic emulsions.

About the Authors

Terence Hetzel
After his studies on instrumental analytics and laboratory management Terence Hetzel started his PhD work at the Institute for Energy- and Environmental Technology (IUTA) in Duisburg, Germany. His research focus is the development and characterization of miniaturized liquid chromatography-based separation techniques in combination with mass-spectrometric detection.

Andreas Hofmann
Andreas Hofmann studied biochemistry at Eberhard Karls University in Tübingen (Germany), University of Massachusetts (USA), and Max Planck Institute of Immunobiology (Germany). He obtained his PhD from the Swiss Federal Institute of Technology (Switzerland) and joined the Novartis Institutes for BioMedical Research (Switzerland) as a laboratory head in 2010. His research focuses on the quantification of peptides and lipids by LC/MS to support early drug discovery projects.

Stavros Kromidas
Stavros Kromidas studied biology and chemistry at the University of Saarbrücken, where in 1983 he obtained his Ph.D. degree on the development of new chiral stationary phases for HPLC. After working for Waters GmbH for five years, he founded NOVIA GmbH, a provider of professional training and consulting in analytical chemistry, serving as the CEO until 2001. Since 2001 he works as an independent consultant for analytical chemistry, based in Blieskastel (Germany). For more than 20 years he has regularly held lectures and training courses on HPLC and has authored numerous articles and several books on various aspects of chromatography.

Friedrich Mandel
Areas of study have included physical chemistry at the Universities of Konstanz and Zurich, several years in the field of doping analysis at the Institute of Biochemistry at the German Sport University Cologne, 1987–2016 application specialist in mass spectrometry at Hewlett–Packard and Agilent Technologies. Application chemist in the fields of GC/MS, LC/MS and MALDI-TOF. Specialized in proteomics and metabolomics, also in training and consulting. Long-time member of the German Chemical Society, DGMS, DGPF, ASMS, ABRF, and SPS. Priorities/topics: LC/MS ionization principles and applications, fundamentals of mass spectrometry.

About the Authors

Markus Martin
Markus M. Martin works as a Product Manager LC Systems at Thermo Fisher Scientific in Germering (Germany). He joined the former Dionex Corporation, now part of Thermo Fisher Scientific, in 2010 as Solutions Manager for LC/MS, being responsible for UHPLC and LC/MS solutions marketing. He received his doctorate in analytical chemistry from Saarland University in Saarbruecken in 2004 for capillary electrophoresis investigations on polyelectrolytes. Before working at Thermo Fisher Scientific, he worked as Analytical Lab Head at Sanofi-Aventis and as Research Fellow at the Saarland University; his scientific work has been focused on UHPLC, HPLC/MS, CE, and CE-MS techniques as well as integrated sample preparation.

Alban Muller
After studying chemistry Alban Muller obtained his Master of Sciences in Analytical Techniques at the University of Strasbourg in France in 2005.
In 2006, he joined Novartis Institute of Biomedical Research in Basel, Switzerland. At that time he was in charge of the ASI (Analytical Sciences and Imaging) platform of nontargeted metabolomics studies by coupling HPLC with high resolution mass spectrometry. Over time, the needs changed to targeted analysis, still under metabolomics activities, and Muller was in charge of looking at small and polar metabolites. Therefore, he started using ion chromatography in 2008 in order to be able to separate and detect small sized and polar compounds.

Oliver Müller
Training as a laboratory assistant (RWE-DEA). Studied chemical/pharmaceutical technology with a focus on analytical chemistry. Research Associate at 4SC AG. LC/MS Field Service Engineer at Waters GmbH. Partner at Fischer Analytics GmbH.

Christoph Portner
After his apprenticeship as a laboratory assistant Christoph Portner studied water science at the University of Duisburg-Essen, Germany. He completed his PhD at the Carl von Ossietzky University in Oldenburg, Germany, in the field of analytical chemistry on the development of LC-MS methods for the detection of mycotoxins. At the Institute of Energy and Environmental Technology (IUTA) in Duisburg, Germany, his research focus was on the identification and quantification of micropollutants, transformation products and metabolites by LC-MS/MS and LC-HRMS. Since 2016 he works at Tauw GmbH in Moers, Germany.

Detlev Schleuder
Detlev Schleuder studied chemistry at the University of Muenster, North Rhine–Westphalia. Completed his PhD thesis in July 2000 in the working group of Professor Franz Hillenkamp, Institute for Medical Physics and Biophysics in the field of MALDI and ESI. Since 2000 at AB SCIEX Germany GmbH (ex. Applied Biosystems). Responsible for the demo lab at SCIEX in Darmstadt and responsible for supporting the food & environmental group in EMEA.

Oliver J. Schmitz
In 2013 Oliver J. Schmitz became a full professor at the University of Duisburg–Essen and is the chair of Applied Analytical Chemistry. In 2009, he cofounded the company iGenTraX UG which develops new ion sources and units to couple separation techniques with mass spectrometers. The research fields of Prof. Schmitz are the development of ion sources, use and optimization of comprehensive LC and GC and coupling analytical techniques with mass spectrometers. He was awarded the Gerhard-Hesse Prize for chromatography in 2013.

Terry Sheehan
After receiving his PhD in Analytical Chemistry from the University of Georgia, Dr. Sheehan's interest in chromatography and mass spectrometry was piqued during a 3-year commitment to the US Army as a clinical chemist/toxicologist at Brooke Army Medical Center. His knowledge was further expanded across four decades of employment with Exxon Chemicals, Varian Associates, Dionex and Agilent Technologies. Within the scope of instrument development and scientific marketing, he gained exposure to many different GC/MS and LC/MS applications including toxicology, clinical, petrochemicals, pharmaceuticals, environmental, and biotechnology. He strongly supports the marriage of chromatography and mass spectrometry and believes the future is bright for both LC/MS and GC/MS with more technological breakthroughs on the horizon.

Thorsten Teutenberg
Thorsten Teutenberg studied chemistry at the University of Bochum, Germany, and received his PhD in high-temperature liquid chromatography in the Department of Analytical Chemistry. Since 2004, he works at the Institute of Energy and Environmental Technology (IUTA) in Duisburg, Germany, where he heads the Division of Research Analytics since 2012. His research is mainly focused on high-temperature LC, miniaturized separation and detection techniques, and multidimensional chromatographic methods.

Jochen Tuerk

Jochen Tuerk studied chemistry at the University of Duisburg–Essen, Germany, and received his PhD on LC/MS-MS method development for pharmaceuticals. At the Institute of Energy and Environmental Technology (IUTA), where he has been division head since 2009, he is concerned with occupational safety, environment, trace and pharmaceutical analytics with LC/MS. His main research focus is on trace analytics and substance identification with LC/MS-MS and LC-HRMS.

Steffen Wiese

Steffen Wiese studied instrumental analytics and laboratory management and received his PhD at the University of Duisburg–Essen, Germany. Since 2007, he is a researcher at the Institute of Energy and Environmental Technology (IUTA). The main focus of his work is the computer-assisted development of methods in liquid chromatography as well as the development of separation, coupling and detection techniques such as high-temperature LC, capillary and nano-HPLC.

Index

a

acetate 48
acetic acid 48, 97
acetone 9, 194
acetonitrile 9, 82, 87, 89, 95, 148, 193
acid
– acetic 48, 97
– carboxylic 63
– formic 48, 49, 82, 97
– organic 48
– trifluoroacetic 48, 49
additive 148
Agent Report Software 85
Agilent system 13, 190
alcohols 63
aldehyde 63
alkali
– adduct 158
– cation 150
alkaline cleaner 194
alkanesulfonic acid 151
all ion fragmentation (AIF) 128
ambient desorption ionization technique 15
amide phase 89
ammonium
– adduct 60
– fluoride 166
– salt 48, 54
analyte molecule 7, 145
analyzer 191
andepirubicin 87
anion exchange chromatography 173
anisole 9
aqueous dopamine hydrochloride solution 187
at column dilution method 180

atmospheric pressure
– chemical ionization (APCI) 6, 9, 44, 143, 171
– – additives 153
– – detection sensitivity 155
– ion sources 3
– – artifact formation 3
– – ion suppression 3
– ionization (API) 5, 142, 159, 177
– – methods 5
– laser ionization (APLI) 5, 10
– photoionization (APPI) 5, 9, 143, 147, 160, 172
– – detection sensitivity 155
– – dopant-assisted 9
autosampler 23
axial mixing 22

b

base 48
baseline stability 51
bicarbonate 48
bioanalysis 49
biomolecule 49
biopolymer 144
biphenyl phase 89
bis-nalbuphine 184
black box 142
bleeding of a separation column 63
blind gradient 53
buffer
– additive 148
– calculator 98

c

calibrant infusion 27
capillary 191
– voltage 155
– flow cell 42

The HPLC-MS Handbook for Practitioners, First Edition. Edited by S. Kromidas.
© 2017 WILEY-VCH Verlag GmbH & Co. KGaA. Published 2017 by WILEY-VCH Verlag GmbH & Co. KGaA.

caprolactam 183
carboxylic acid 63
cation adduct 159
charged
– aerosol detectors (CAD) 40
– residue mechanism (CRM) 8
charging effect 193
chemical noise 56
chip-design mixer 22
chloride
– ion 48
– salt 61
chromatogram 23, 28
– baseline quality 56
– divided into periods 125
chromatographic support 100
chromatography data systems (CDS) 217
Chromeleon 218
Chromolith 179
circular ion trap (QIT) 68
citric acid 186
classroom training 166
cleaning 193
cloud computing 218
collision
– cell 208
– energy 192
collision-induced dissociation (CID) 34, 158
column
– geometric volume 106
– inner diameter reduction 132
– reequilibration 31
– switching valve 102
– thermostat 26
– volume 107
cone 191
– voltage 155, 166, 190
connection tubing 26
constant neutral loss 35
contamination 63, 162
conversion
– detector 193
– dynode detector 198
coordination ionspray 6
core–shell particle 100
costs 205
Coulomb explosion 144
critical radius 8
CSH-Xselect column 179
curtain gas 116, 207
cycle time 23, 30, 122, 127
cyclophosphamide 81, 87, 89
cytostatic drugs 119

d
data
– acquisition rate 38
– management system 217
database 130
data-dependent acquisition (DDA) 128
data-independent acquisition (DIA) 128
declustering 58
decomposition product 177, 181
deconvolution 205
degradation experiment 181
delayed injection 30
derivatization 203
design of experiments (DoE) 118
desolvation 142
detection
– selectivity 204
– sensitivity 153, 204
– technique, concentration-dependent 144
– window 126
detector
– electrochemical 42
– nebulizer-based 42
– nondestructive 42
dewetting 84
dimethylbenzidine 156
dimethylformamide 148
DIN 38407-47 75
diode array detection 41
direct injection 104
display mass 177
dissociative electron-capture ionization 6
dissolved residual gas 65
diverter valve 162
docetaxel 81, 87, 120
dopamine 186
– impurities 186
dopant 9, 147
dopant-assisted atmospheric pressure photo ionization (DA-APPI) 9
double peak 106
doxorubicin 81, 87
drug screening 128
dryer gas 117
Drylab 98
dwell time 122, 124

e
electro spray ionization 171
electrochemical detector (ECD) 42
electrochemistry 40
electrodynamic ion funnel technology 172
electron
– impact ionization 62

– multiplier 193
– – tube 197
electrospray 190
– ionization (ESI) 3, 5, 12, 44, 104, 142, 177, 207
– – additives 152
– – detection sensitivity 154
– – ESI needle 116
– – ESI voltage 58, 117
– – positive ion ESI 159
eluent mixing 22
enoxaparin 177
environmental analysis 73, 77, 211
epirubicin 81, 87
equilibration 85
etoposide 120
evaporation enthalpy 47
evaporative light scattering detector (ELSD) 40, 177
extra-column volume (ECV) 20, 21, 25
extracted ion chromatogram (EIC) 41, 57

f
false mass
– assignment 65
– determination 68
fine optimization 98
fitting technology 28, 214
flow
– injection analysis (FIA) 45, 50, 57, 118, 158
– rate 103
flow-through-needle principle 23
5-fluorouracil 81, 87, 109
flush-out effect 21
foamtips 197
formethotrexate 97
formic acid 48, 49, 82, 97, 194
fragment ion 34
fragmentation reaction 61
fragmentor voltage 166
freeze-drying 187
FTICR mass spectrometer 36
full-scan
– mode 209
– spectrum 129
fully porous particle 100
gas
– filter 57
– chromatography (GC) 203
– phase 43
– – adduct 59, 60
– purity grade 57

– stream 117
gemcitabine 81, 87, 109
gradient
– delay volume (GDV) 20, 21, 24, 215
– generator 174
– grade solvent 182
– separation 52
– slope 95

h
Hagen–Poiseuille's law 25
heptanesulfonic acid 151
hexapole 193, 198
high resolution analysis 31
high-pressure gradient pump (HPG) 22
high-resolution/accurate mass (HRAM) 215
high-throughput analysis 20
hot steam cleaner 194
HPLC
– analysis 20
– flow rate 154
HPLC/MS
– analysis 75
– coupling 77
– – critical peak pairs 77
hybrid
– mass spectrometer 75
– system 209
– – QTOF architecture 210
hydroniumion 174
hydrophilic liquid interaction chromatography (HILIC) 46
hydroxyeicosatetraenoic acid 173
Hypercarb 109
– precolumn 110

i
ifosfamide 81, 87, 89
impurity 182
– on a gram scale 183
inappropriate mass resolution 66
inert surface 194
information-dependent acquisition (IDA) 128, 209
injection volume 104, 108
inner column diameter 103
in-source collision-induced dissociation 61
ion
– block 198
– bridge 191
– burn 196
– chromatography (IC) 173, 174
– cyclotron resonance (FTICR) 216
– – mass analyzer 35

– evaporation mechanism (IEM) 8
– mobility 13
– – mass spectrometry (IMS) 13, 15, 205
– optic 162
– pair
– – agent 160
– – chromatography 152
– – reagent 150, 160, 173
– source 3, 43, 150, 162, 190
– – chemical contamination 56
– – inappropriate settings 56
– – parameters 51
– – sprayer 27
– suppression 48, 53, 58, 122, 158, 162
– – by the sample matrix/sample contaminants 161
– – determination 11
– transfer
– – capillary
– – optic 44
– trap 172
– – mass spectrometer 12, 34, 36
– – Orbitrap 172
ionization
– marker 10
– method 5
– optimization 159
– potential 9
– techniques 142
– voltage 117
IonSpray 144
irinotecan 108
isomer 40
isopropanol 109, 194, 197
IT application 218

l

lab information management systems (LIMS) 219
Lambert–Beer's law 155
large molecule analysis 215
large volume injection (LVI), see direct injection
LC/MS
– coupling 3
– – method development 43
– – problem solving 177
– – technical aspects 19
– interface 142
– separation setup 53
Leu-enkephalin 47
light scattering detection 177
limits
– of detection (LOD) 104

– of quantification (LOQ) 104
linear
– ion trap (LIT) 68
– solvent strength (LSS) model 109
linearity 156
liquid chromatography (LC) 6, 19, 49, 203, 207, 214
lock spray 36
low-pressure gradient pump (LPG) 22
LTQ Orbitrapmass spectrometer 13

m

maintenance engineer 189
make-up flow 101
mass
– resolution, inappropriate 66
– signal
– – misinterpretation 66
– – unknown 59
– spectrometer 12, 33
– – cycle times 38
– – data rates 38
– – high-resolution 78
– – source parameters optimization 117 117
– spectrometry (MS) 5, 19, 215
– – contaminant database 65
– – control software 68
– – detectability 39
– – detection parameters 51
– – establishing parameters 115
– – hybrid instruments 172
– – method development 73
– – tuning 50
– spectrum misinterpretation 65
– transition 78
mass-selective detector (MSD) 37
matrix
– blank 164
– effect 52, 162, 165
– – chromatogram 79
– spike 163
matrix-assisted laser desorption ionization (MALDI) 177
membrane-based pulsation damper 23
metabolomics 174
metal ion 64
meter column length 214
metering device 24
methanesulfonic acid 173
methanol 89, 92, 182, 193, 197
method scouting system 84
methohexital 181

microdroplet 7
microspray 144
microwave vials 184
miniaturization 132
mixed-mode phase 89
mobile phase flow rate 103
monolithic phase 100, 102, 133
mother drop 7
multianalyte method 79, 104
multiple reaction monitoring (MRM) 34, 122, 183, 157
– retention-time-dependent 125
– retention-time-independent 124
– scheduled (sMRM) 208
myoglobin 145

n

nalbuphine 184
nano-HPLC technique 142
nano-LC system 51, 171, 207
nanospray 144
nebulizer gas 8, 117, 120
– improper 58
negative ion electrospray 160
nitrogen 56, 190, 191
– generator 57
non-target
– analysis 13, 211
– screening 74
nozzle-skimmer dissociation 62

o

offline SPE 114
oligomer separation 183
oligonucleotide 166
online SPE 112
Orbitrap 13, 36, 75, 129, 213, 215
organic
– acid 48
– solvent 89
overspray 190
oxidation
– electrochemical in the ESI ion source 61
– photochemical 61

p

paclitaxel 81, 87, 120
PAH analysis 11
paired ions 150
pause time 124
peak
– area calculation 38
– compression 24, 29
– width 127

PEEK fingertight fitting 28
PEEKSil capillary 132
pentapeptide 47
perfluorinated
– compounds 63
– organic acid (PFOA) 63
pesticide screening 99, 128
pH value 95, 160
pharmaceutical selection 80
phase collapse 84
phosphate 48
– buffer 54, 148, 149, 193
photoionization (PI) 6, 146
– detector (PID) 6
photomultiplier 193, 198
photoreactor 61
plasticizer 63
pneumatic nebulizer 144
– gas 44
pneumatically assisted electrospray 144
polarity switching 123
polyether 64
polyethylene/polypropylene glycol (PEG/PPG) 64
polyfluorinated compounds 63
polysiloxane (silicone) 64
porous graphitic carbon (PGC) 109
post-column
– addition 152, 153
– infusion 58
potassium hydroxide 173
precursor ion 34
– scan 35
prednisolone 152
prefilter 192
proteomics 144, 171
proton adduct 59
pseudomolecular ion 142, 153
PTFE (Teflon) 63

q

QTOF system 15
QTRAP technology 209
quadrupole 172, 191
– cleaning 195
– mass spectrometer 12, 36
– Orbitrap 173
– time-of-flight mass spectrometer 75, 129
qualifier 208
quality
– by design (QbD) 118
– control (QC) analysis 53
quantification 122
quantifier 208

r

radial mixing 22
reagent gas 45, 148
re-equilibration time 31
remote device monitoring 218
residue analysis 122
resolution map 99
resonance-enhanced multiphoton ionization (REMPI) 10
response factor 44
retention time 80
– window 125
reversed-phase (RP) chromatography 46
RP chromatography 40, 82, 173

s

saffron sample 4
sample
– analysis 54
– – no signal 54
– injection 23, 30
– loop 112
sandwich injection 180
SCIEX (SCIentific EXchange) 207
screening
– experiment 84
– for unknowns 34, 35
– fully automated 84
– manual/partially automated 85
– using LC/MS 128
selected reaction monitoring (SRM) 34
sensitivity 44
sheath gas 8
signal
– intensity 55
– quality 51
– reduction 58
signal-to-noise ratio 99
simulation software 98
single ion mode (SIM) 12
single reaction monitoring (SRM) 157
size-exclusion chromatography (SEC) 92
skimmer 191, 194
small molecule analysis 49, 215
SmartInject technology 214
soft clipping phenomenon 156
software 217
– tools 119
solid phase extraction column 105
solvent 148
source temperature 117, 119
space charge 68
spectral library 141
spectroscopy 40
– detector 38
split-loop sampler 23, 30
spray
– stability 50
– techniques 190
stainless steel (SST)
– capillary 29
– cleaner 194
stationary phase support 99
supercritical fluid chromatography (SFC) 14
suspected-target screening 74
syringe pump 51
system tubing 24

t

tandem
– liquid chromatography 32
– MS in space 34, 35
target
– analysis 15, 73, 74, 122, 208
– scan time 126
– screening 34, 37
Taylor cone 7
temperature 92
tetrabutylammonium 151
– bromide 151
tetrahydrofuran (THF) 92, 148
thermal spray 190
Thermo Fisher Scientific 213
ThermoScientific EasySpray nano-LC interface 217
time window 124
time-of-flight (TOF)
– analyzer 210
– mass spectrometer (TOFMS) 13, 35, 38, 129, 145, 172, 215
toluene 9
total cycle time 30, 33
total ion chromatogram (TIC) 57, 129
trace analysis 122
transfer capillary 100, 144
triethylamine (TEA) 149
trifluoroacetic acid (TFA) 48, 49, 149, 173
triple quadrupole mass spectrometer 12, 34, 38, 78, 122, 207
TripleTOF technology 211
tuning solution 51

u

UHPLC
– analysis 20
– fitting system 28

– system 127, 214
– – installation 25
ultrasonic bath 194
uniform
– collision energy 130
– detection 39
unit mass resolution mass spectrometry 204
universal detection 39
UV detector 27, 38, 42, 55, 61, 77

v
vacuum
– degasser 152

– ultraviolet (VUV) radiation 6
van Deemter curve 46
Vanquish UHPLC system 24
volatility 47

w
waste valve 100
Waters system 13
wide area networks (WAN) 217

z
Z-spray 190